Cost Reduction
and Optimization
for
Manufacturing and
Industrial Companies

Scrivener Publishing
3 Winter Street, Suite 3
Salem, MA 01970

Scrivener Publishing Collections Editors

James E. R. Couper	Richard Erdlac
Rafiq Islam	Pradip Khaladkar
Norman Lieberman	Peter Martin
W. Kent Muhlbauer	Andrew Y. C. Nee
S. A. Sherif	James G. Speight

Publishers at Scrivener
Martin Scrivener (martin@scrivenerpublishing.com)
Phillip Carmical (pcarmical@scrivenerpublishing.com)

Cost Reduction and Optimization for Manufacturing and Industrial Companies

Joseph Berk

Scrivener

Co-published by John Wiley & Sons, Inc. Hoboken, New Jersey, and Scrivener Publishing LLC, Salem, Massachusetts
Published simultaneously in Canada

For general information on our other products and services or for technical support, please contact our Customer Care Department within the United States at (800) 762-2974, outside the United States at (317) 572-3993 or fax (317) 572-4002.

Wiley also publishes its books in a variety of electronic formats. Some content that appears in print may not be available in electronic formats. For more information about Wiley products, visit our web site at www.wiley.com.

For more information about Scrivener products please visit www.scrivenerpublishing.com.

Cover design by Russell Richardson.

Library of Congress Cataloging-in-Publication Data:

ISBN 978-0-470-609576

Printed in the United States of America

10 9 8 7 6 5 4 3 2 1

Dedication
and Acknowledgments

I HAVE WORKED WITH great manufacturing leaders. This book is dedicated to Johnnie Crean, Chuck Sebastian, Ed Elko, Irv Barger, Greg Burns, and John Gozza. Thank you for all you have taught me.

PneuDraulics, Incorporated, a world-class developer and manufacturer of precision hydraulic aerospace components in Rancho Cucamonga, California, allowed access to their operation for many of the management concepts and photographs in this book. PneuDraulics is a great company, and I wish to thank their employees and the management team for their assistance.

Contents

Introduction

Every business can reduce its costs. The need for doing so is even more urgent in times of recession, but cost reduction is always important. It shouldn't take failing banks, tightened credit, and federal bailouts to make people realize that keeping costs down is important. However, in many cases when times are good, managers get sloppy. Sometimes the money is rolling in so fast that cost reduction seems to be irrelevant. The danger is that times don't stay good, and if an organization doesn't make cost reduction a priority in good times and bad, it will be left in the dust when times turn bad.

Every manufacturer and every industrial company wants to reduce costs, but knowing how to do so without adversely affecting quality or client satisfaction can be difficult. Cost reduction is not taught in business schools, engineering programs, or any other academic environment. Cost reduction is not an innate skill for most of us, but it can be learned. It's not that difficult.

I started focusing on how to reduce costs in a large manufacturing organization many years ago. I worked for a visionary leader who had accepted a quarter-billion dollar production contract at a projected loss. He wanted to turn that projected loss into a profit. That was my job. I was charged with reducing costs enough to turn a contract with a projected loss into one with a healthy profit.

I worked with brilliant and creative engineers, manufacturing managers, and purchasing specialists. We were successful in dropping costs in every area. After doing so, I analyzed and categorized what we did, and I realized two things:

- This isn't rocket science. For the most part, cost-reduction actions are commonsense endeavors.
- The previous statement notwithstanding, there is a technology here – a cost-reduction technology.

I grew that experience into a successful consulting practice helping other companies lower their costs and become more profitable. *Cost Reduction and*

Optimization for Manufacturing and Industrial Companies is based on practices developed in that first cost-reduction assignment and on refinements developed in the next twenty years of consulting engagements with hundreds of organizations.

I tell you these facts so that you know my background. It's probably not too different from yours, and I will tell you up front that I'm not any smarter than you are. What I bring to the table is that I have focused on this area for nearly a quarter of a century. That experience is here for you to use.

Cost Reduction and Optimization for Manufacturing and Industrial Companies is intended to be an easy book to read. I wrote it for supervisors, managers, and executives. It includes easy-to-understand and easy-to-implement cost-reduction concepts. The book has six general areas: labor, material, design, process, overhead, and measuring progress. The chapters are simple but direct. Each chapter:

- Dives into a cost-reduction area and starts with the bottom line first by summarizing key points.
- Asks a few questions to help you determine if the area presents an opportunity for your organization.
- Provides a road map for implementing recommended actions.
- Presents information that completely and succinctly explains the relevant concepts.
- Identifies who in the organization should do the work.
- Outlines risks and suggested risk-mitigation actions.

That last point is critical because cost reduction involves change, and wherever change is present, so are risks. You have to know what the risks are in order to manage them.

You can read this book from front to back in a day, or each chapter can be used in a stand-alone manner. The chapters are short. Each chapter provides proven tactics for cutting costs in a specific area. Some topics are covered in a few pages and others take longer, but the objective in each chapter is to get to the point without a lot of extraneous data. Where mathematics and other more complex tracking methodologies are needed, there are instructions on how to use Excel to make it easy. There are a lot of tables, graphs, and photos to show the concepts graphically because that makes them easier to understand.

As an added bonus, we are making an Excel file available to *Cost Reduction and Optimization for Manufacturing and Industrial Companies* readers for free (the file is available at www.ManufacturingTraining.com). You don't have to download the free Excel file to use this book (the concepts and the approach are explained fully in these pages), but it can save you a bit of time in areas where spreadsheets can accelerate the analysis, implementation, and management of each concept.

1

Organizing a Cost-Reduction Program

The Bottom Line

You need a multidisciplinary team to attain significant cost reduction. Support from the top helps greatly. You will encounter resistance to the cost-reduction effort and there are risks associated with cost-reduction activities, but these issues can be overcome. The team needs to prioritize cost-reduction opportunities, assess the necessity of all costs, quantify projected savings, identify implementation costs and risks for each proposed action, meet at least once a week and maintain an action plan to create and sustain cost-reduction momentum.

Key Questions

Do we have a cost-reduction effort in place?

Do we have cost-reduction targets?

How do we identify and eliminate unnecessary costs?

What obstacles will we encounter, and how will we get around them?

The Cost-Reduction Program Road Map

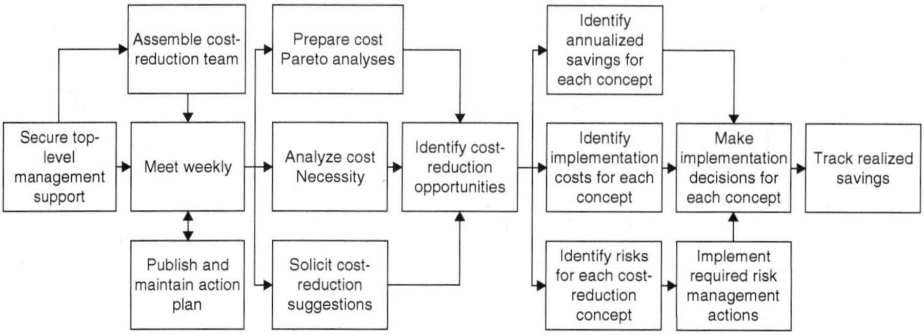

Figure 1.1 The cost-reduction road map.

Teamwork

If you want to reduce costs in your company, you can't do it by yourself. There are cost-reduction opportunities in every department. Identifying and implementing these cost reductions requires the enthusiastic cooperation of people in sales, finance, engineering, manufacturing, quality assurance, purchasing, facilities, and human resources. Even if you wish to limit cost reductions to a single area, you'll still need help from the people in that area and probably the finance organization. You can't mandate cost reduction. You have to have help from the people who will make it happen.

Senior management support will help to make the cost-reduction effort successful. If your interest in cost reduction is the result of a directive from the organization's chief executive, you already have the senior-level support you need. If your effort is self-initiated, support from the person at the top is a great asset. You need support from other cost-reduction team members, but if the chief executive is on board, others will be more enthusiastic about supporting the effort.

The best way to identify and implement cost reductions is to build a team with one or two people from each area who believe in the mission. This team should be made up of people who are already in the company. You don't need to hire more people for this (in fact, a recurring theme throughout this book will be to keep the headcount as low as possible).

Resistance to Change

Most people are naturally resistant to change, and cost reduction will involve change (sometimes big change). Department managers and others may resist cost-reduction–related changes for any of several reasons:

- The idea was not theirs.
- The idea will involve effort on their part.
- They did not think of the idea first, and perhaps that is a source of embarrassment.
- The idea has implementation and operational risks.
- There may be turf issues, where the team is recommending eliminating or modifying a pet project, or the affected managers don't like the idea of someone else suggesting how their departments should operate.

All of these resistance-to-change factors are likely to be encountered as the cost-reduction effort proceeds. All of these arguments must be overcome if the cost-reduction effort is to succeed. It's a lot easier for people to accept change if the general manager or company president is visibly and consistently behind it. That's not the only requirement for overcoming resistance to change, but without top-level support, it will be harder to overcome.

The Cost-Reduction Team

If the chief executive asks you to head up a cost-reduction effort, you are in a good position. The most important thing you should ask for is that you get good people on the cost-reduction team. You don't necessarily want the head of each functional area, and you certainly don't want people who are less-valuable employees within their departments. You want people who:

- Are bright, curious, and "out of the box" thinkers.
- Have a high energy level.
- Make things happen.
- Meet their schedule commitments.

If you can assemble a team with people meeting these criteria, you are going to have a lot of fun and your company will realize great savings.

In the first meeting, the first assignments should be *identifying and ranking the organization's current costs, and assessing the necessity of each cost*. You won't be able to do all of this in the first meeting, but the team members should be able to have gathered this information by the next meeting. This will require support from the finance department (as will many cost-reduction activities, which is why it makes sense to have a finance person on the team).

The cost reduction team should meet weekly at a minimum, because if you meet less frequently the effort will lose momentum. Here's a suggested approach for how the cost-reduction meetings should be run:

- You (or someone else who writes well and is good at capturing details) should take notes and publish meeting minutes no later than one day after each meeting. The meeting minutes should be sent to the chief executive, the team members, and the heads of each department. Doing this keeps others in the loop, and it keeps the effort alive.

- The meeting minutes should include a "living" task list. We'll present a suggested format and say more about this in a bit. Team members should provide input regarding the status of each task in the meeting, and the person preparing the meeting minutes should update the task list to show current status.

- The team members should discuss cost-reduction ideas in a free-flowing manner. The ideas may come from the team members or from others in the company. All of the ideas should be captured on paper. After discussing all of the ideas, the team should decide if each idea should be pursued. If the team thinks an idea has merit, in most cases it will go to the affected department manager. We'll talk more about this later.

Cost Pareto Analysis

Identifying and ranking all of the organization's current costs is best presented on a department-by-department basis, and by overhead cost categories for the entire company. We recommend presenting this in a Pareto[1] format. It's important to do this for each department and by overheard cost category to identify where the greatest opportunities exist. Within the manufacturing area, for example, labor and material costs are probably higher than other costs, and based on that, they probably have greater cost-reduction opportunities. Smaller cost categories will also offer opportunities (and there may be low-hanging fruit that the team wants to grab), but in general the larger cost categories offer greater opportunities when seeking cost reductions.

Let's assume the manufacturing department reviews its monthly operating costs when they receive them from the finance department, and they find the following:

---

1. Vilfredo Pareto (1848–1923) was an Italian economist who is credited with originating the 80-20 rule when he observed that 80 percent of Italy's wealth was concentrated in 20 percent of the population. This led to the creation of the 80-20 concept and Pareto charts, which show most-frequently-occurring to least-frequently-occurring items, or most costly to least costly expenses. The idea is that efforts should be focused on the most significant areas.

With this information in an Excel spreadsheet, it is a simple matter to sort the data (it's already been sorted in Table 1.1) and prepare the Pareto chart shown in Figure 1.2).

Table 1.1 July Manufacturing Department Costs

Cost Category	Cost
TB Steel	$227,950
Labor	$188,160
Paint	$66,560
Supervision	$54,000
Overtime	$50,400
Maintenance	$18,992
Tooling	$14,777
Electricity	$13,562
Weld gas	$7,285
Supervisor car leases	$7,012
Fuel	$6,783
Weld Rod	$5,934
Travel	$4,254
Coffee	$3,760
Training	$3,250

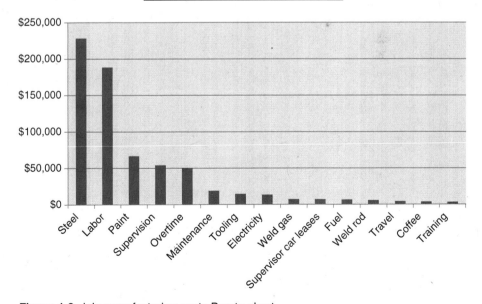

Figure 1.2 July manufacturing costs Pareto chart.

Based on the information presented in Figure 1.2, it is obvious that material and labor are the largest cost categories. Logic dictates that seeking cost-reduction opportunities in these areas offers the best potential. In addition to steel and labor costs, overtime (a frequently abused area) pops out as a relatively large cost, so it should also become a cost-reduction target.

Once the analysis has been completed for each department and for the company's overhead costs, the team can then brainstorm reduction activities in these areas. The team can also apply the techniques to be reviewed in detail throughout this book.

As mentioned earlier, the team's activities should not be limited to just the largest cost categories. We're only suggesting that because of their size, these "big hitters" probably contain greater cost-reduction opportunities. There will be opportunities in the lower cost areas that come from other people's suggestions as well as the cost-reduction team members. The team should consider these as well.

Assessing Necessity

The next action is identifying the magnitude and evaluating the necessity of each cost item. This is best done on a department-by-department[2] basis and for overhead costs. Table 1.2 shows this for the manufacturing area.

The idea here is to identify each cost as necessary, unnecessary, or nice-to-have. The team may wish to use different necessity descriptors, but the concept is to identify unnecessary costs and nice-to-have items as potential candidates for elimination. Necessary items should not be eliminated, but they may be candidates for further analysis using the techniques described in the rest of this book.

The first part of this task is relatively easy (identifying the cost items). The second part (evaluating the necessity of each) is much more subjective. Resistance to cost reductions in some of these areas will almost certainly emerge. Sometimes when costs are presented this way, the frivolity of the unnecessary costs becomes obvious and no resistance occurs.

Table 1.2 shows a recommended approach for accomplishing this.

---

2. One can argue that the department managers and supervisors should be making these assessments as an ongoing part of their jobs. Although this is true, it frequently does not occur. The exercise described here lends rigor to the effort.

Table 1.2 Cost Necessity Assessment

Cost Driver	Annualized Cost	Classification	Risks	Decision
Steel	$2,358,992	Necessary	None	Keep, but reduce cost
Labor	$2,257,920	Necessary	None	Keep, but reduce cost
Paint	$851,400	Necessary	None	Keep, but reduce cost
Supervision	$648,000	Necessary	None	Keep
Overtime	$645,000	Necessary	Morale impact	Keep, but reduce cost
Maintenance	$265,888	Necessary	None	Keep, but explore outsourcing
Tooling	$192,101	Necessary	None	Keep, but reduce cost
Electricity	$135,620	Necessary	None	Keep, but reduce cost
Weld gas	$88,149	Necessary	None	Keep
Supervisor car leases	$84,144	Nice to have	Morale impact	Evaluate other options
Fuel	$81,396	Necessary	None	Keep, but reduce cost
Weld rod	$77,142	Necessary	None	Keep
Travel	$55,302	Necessary	None	Keep
Coffee	$45,120	Unnecessary	Morale impact	Eliminate
Training	$32,500	Nice to have	None	Keep, but reduce cost
Department party	$6,300	Unnecessary	Morale impact	Keep
Supervisor golf club memberships	$5,000	Unnecessary	Morale impact	Eliminate

Cost-Reduction Action Plans

Maintaining and updating an action item list is critical. It assures that each cost-reduction concept is captured on paper and retained until it has been objectively evaluated, and either discarded or implemented. It's also important because it identifies who needs to do what, by when they need to do it, and current status.

Table 1.3 Recommended Cost-Reduction Team Task Action Plan Format

Concept Number	Concept	Required Actions	Assignee	Required Completion Date	Implementation Cost	Estimated Annualized Savings	Risks	Status
1	Reduce steel costs	Determine if steel drop-off can be reduced	Smith	15 Sep	$8,000	$225,000	None	In work
		Introduce supplier competition	Gonzales	30 Oct	$4,000	$50,000	Offending current supplier	Not started yet
2	Reduce labor costs	Develop and implement labor standards	Nguyen	30 Nov	$20,000	$125,000	None	In work
		Develop and implement efficiency measurement system	Nguyen	30 Dec	$5,000	$125,000	Morale; inaccurate standards	Not started yet
		Flowchart production process to identify cost-reduction opportunities	Jackson	15 Aug	$1,000	TBD	None	Complete
3	Reduce paint costs	Develop and implement paint application methods training	Aker	10 Aug	TBD, considered to be low	$160,000	None	In work; behind schedule
		Consult paint suppliers to identify paint usage–reduction techniques	Aker	30 Aug	$10,000.	$100,000	None	Complete

#	Goal	Action	Responsible	Date			Risk	Status
4	Reduce overtime	Identify overtime causes	Thomson	15 Sep	$1,000	See below	None	Complete
		Develop and implement overtime budget	Dept Heads	20 Sep	$0	$15,000	Morale; making sure work is complete	Not started yet
		Develop and implement overtime request form	Thomson	20 Sep	$0	$8,000	Department manager's acceptance	Not started yet
5	Eliminate free coffee	Secure approval from chief executive	Spitler	15 Aug	$0	See below	See below	Complete
		Notify all personnel and implement	Dept Heads	15 Aug	$0	$45,120	Morale	Implemented
6	Eliminate golf club memberships	Discuss issue with chief executive and department heads	Jones	15 Aug	$0	See below	None	Not done yet; behind schedule
		If decision is to proceed, notify supervisors	Dept Heads	30 Aug	$0	$5,000	Morale, industry perceptions	See above
7	Reduce maintenance costs	Identify dominant maintenance cost contributors	Balicki	30 Sep	$500	TBD	None	In work
		Determine if more preventive maintenance would lower costs	Balicki	30 Sep	$500	TBD	None	Not started yet
		Explore outsourcing maintenance function and make recommendation	Balicki	30 Sep	$500	TBD	Existing maintenance staff morale	Not started yet

Notes:
Gray-shaded row indicates action complete.
Red-shaded row indicates action behind schedule.

11

The action item list becomes the cost-reduction plan, and maintaining and circulating it lets everyone know how the team is doing.

Table 1.3 shows a format that works well. As the table indicates, completed actions are shaded in gray and actions that are behind schedule are shaded in red. This gives the team and everyone who reviews the plan a quick look at how things are progressing. Nearly every assigned action will require the efforts of more than one person, but it's always a good idea to list only one name for each action in the plan. The chances of the action being completed on schedule are higher if it has a single owner.

Table 1.3 shows the plan for only one area in the organization, and the example shown here is not intended to be complete. The action plan needs to address all areas.

Quantifying Estimated Savings and Implementation Costs

The actions the team identifies in its plan are all intended to produce a cost savings, but before making any changes, the team should objectively estimate what the savings will be. Great care in maintaining objectivity is required here. It is very easy to overestimate planned savings.

Most cost-reduction concepts require implementation actions, and there is usually a cost associated with these actions. Sometimes there are no costs (for example, if an unnecessary step is eliminated), but most of the time there are costs associated with changes. For example:

- Engineers may need to redesign products.
- Manufacturing engineers may need to redesign processes.
- The purchasing department may need to obtain prices from alternative suppliers.
- New production equipment may be required.
- Facilities engineers may need to modify the building.
- New designs may require testing to confirm requirements compliance.
- There may be disruptions as the change is implemented.

These actions all involve cost, and it is important to accurately predict what these costs will be. The obvious reason for doing this is that it makes no sense to implement a cost-reduction change if the implementation cost exceeds the savings.

Our experience indicates that it is best to quantify the savings on an annualized basis, and to consider the implementation costs in the first year. Sometimes in job-shop production environments with production runs lasting less than a year it is best to quantify savings on a contract basis. This is another area where the finance department can help.

Who Should Do This Work

The cost-reduction team can't do everything. In fact, by itself the team won't be able to evaluate or implement most of the ideas. The team is a catalyst. It has to work with others in the organization to gain support, evaluate ideas, and where appropriate, implement cost-reduction actions.

Risks

Risk has to be a key part of the decision process when evaluating every cost-reduction concept. In assessing risk, there are three questions to consider:

- Have we done this before?
- What can go wrong if we do this?
- What are the things we need to do to prevent bad things from happening?

If the concept being considered has been implemented elsewhere, or if it consists of smaller actions that have been done before, the risk is probably minimal. If neither condition exists, the organization needs to aggressively and objectively identify all potential consequences (not just the planned savings) and take steps to manage the risks. Risks may include customer reaction, product performance, supplier reliability, morale, process yield, safety, and other factors. Specific risks in different areas are identified in each chapter throughout the rest of this book. The important thing to recognize is that cost reduction involves change, change carries associated risk, and prudent people identify and manage risk.

References

W.J. Stevenson, Production/Operations Management, New York, McGraw-Hill, 2000.
R.W. Bradford, *Simplified Strategic Planning: The No-Nonsense Guide for Busy People Who Want Results Fast*, Worcester, Chandler House Press, 2000.

Part I

Labor

Labor optimization is a key part of any cost-reduction effort. It's not the only part, but is an important part, so we will address it as the first of our six major areas. This section focuses on:

- Defining an appropriate headcount and controlling staff growth.
- Developing and implementing time standards.
- Measuring and using efficiency to improve performance.
- Assessing machine utilization and improving it.
- Controlling overtime.
- Making multiple shifts productive.
- Finding and reducing lost time.
- Using the learning curve to drive costs down.

2

Defining Headcount
and Controlling Staff Growth

The Bottom Line

Headcount should be based on the organization's required output, the work content, the time required for the work, and reasonable efficiency estimates. The organization can use this objective, quantitative method for determining headcount in both direct labor and overhead areas. Headcount tends to grow if not aggressively managed; the organization should add staff only if the need can be justified after considering all other alternatives.

Key Questions

How many people do we have?

How many people do we need?

How do we determine how many people we need?

Who approves staff additions?

Are there alternatives to hiring more people?

The Headcount Determination Road Map

When cost reduction is mentioned, many people assume that headcount is the first and perhaps only target. Headcount is an important cost element, but it is not the only area to consider when reducing costs. Headcount needs to be addressed, though, because labor is usually a huge part of the cost base. It's easy to add headcount, and headcount tends to increase in organizations not focused on matching headcount to work content. The result is increased cost.

The issue is defining how many people the operation actually needs to operate. Sometimes this question emerges when starting a new facility or product line, and sometimes it emerges when someone wants to add people to an existing area. It should be a question that management considers all the time, even in stable production environments.

Based on our observations, this is what usually occurs when organizations define how many people they need:

- In new organizations, management creates a structure using people from elsewhere in the organization, by hiring people, or by doing both. The headcount determination is often driven by a subjective assessment of what the organization will have to do. Translating the work into headcount is typically based on qualitative reasoning.

- In existing organizations, departments and work centers already exist. Over time, situations and requirements change (new work is accepted,

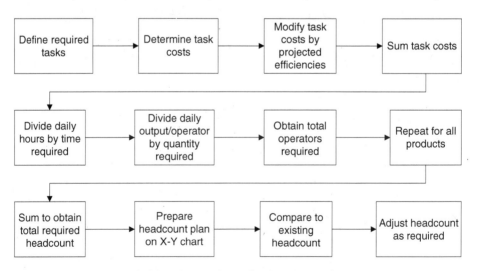

Figure 2.1 Recommended headcount determination approach.

responsibilities grow, new equipment is acquired, people become available from other parts of the organization, etc.). This occurs constantly, but it especially occurs when sales increase. The logic is "We are probably about right now, so if we have more work to do, we must need more people."

In consulting for hundreds of companies, these approaches are the ones we most frequently encounter. These approaches are dangerous. They add expense. They do not base headcount on work content. Managers who use these approaches are not focused on accomplishing the organization's objectives at the lowest possible cost.

Consider an example that highlights the difference between hiring more people simply because it seems like more people are needed versus quantitatively determining work content and using that to drive headcount.

We Need More People

At a trailer manufacturer, a relatively new supervisor was having trouble meeting the production schedule. He asked his manager if he could hire another person. His manager's response was simply, "Okay."

The supervisor was stunned. "Just like that? I don't have to justify it or anything?"

"No," said the manager, "We've already done that. In fact, you can hire as many employees as you want. Are you sure one is enough?"

The supervisor was confused. "I think so," he said. "Why do you say I can hire as many employees as I want?"

"Well," the manager answered, "We have a certain value allocated to your work center, and it's based on the work content, you know, what you guys have to do. We converted that into the number of people in your area, and that's our budget for your area. That's all the money there is, and that's how we came up with the number of employees you have. If you feel you need more employees, though, that's up to you. We'll just divide the new number of employees into what's available for your work center. You can have more employees, but the hourly rate for each one will have to go down to keep the total amount of money we pay you guys the same. It works the other way, too. If you can do the work with fewer people, your hourly rates will go up. That's the way it works here. We've studied your area, and we know how much work is there. That's how we defined the budget for your area."

Later that day, the supervisor visited the manager again. "We don't need another guy," the supervisor said. "We talked about it, and we don't need as many people as we have now. We want to reduce our headcount by one person."

Assessing Requests to Add Staff

Suppose your organization is considering adding more people. Here are a few questions that will quickly reveal if the organization is headcount sensitive, and if it bases staffing levels on quantitative analysis.

How many people do we have now?

This should be an easy question. Amazingly, in many cases supervisors and managers don't know the answer. Obviously, any organization that doesn't know its current headcount is not headcount sensitive. In any organization that doesn't know its current headcount, the analysis supporting the need for more people has to be soft.

Why do we need more people?

This obvious question is a great one. If the answer is simply "because there's more work" with no quantified analysis to convert the added work content into a new headcount, the need for additional people may or may not be real. More analysis is required to determine the appropriate number.

Is the increased workload a permanent situation, or is it transitory?

If the increase is permanent, adding staff may make sense (assuming the analysis behind supporting the request is quantitative and it makes sense). If the increased workload is temporary, it might make more sense to address it through overtime, by subcontracting the work, or perhaps by other approaches explained in the answer to the next question.

What other options have we considered?

If the answer is none, the request for more people has not been adequately thought out. There are many options to hiring more people. These include using overtime, transferring people from other departments, revising the process to improve efficiency and utilization of existing resources, deciding not to do the additional work (a decision admittedly not within the supervisor's purview, but one that needs to be considered at a higher level nonetheless), or subcontracting the work.

How busy are our people now?

The obvious and most frequently heard answer to this question is "very busy," but if the reasoning behind the answer is not supported by quantitative data related to efficiency, capacity, workload, and other factors to be discussed in subsequent chapters, the answer is probably not supportable. These are the bits of information needed to truly support an increased headcount request.

If they're missing in the answer to this question, the logic supporting the need for more people is weak.

How much overtime are we working now?

If the supervisor doesn't know the answer to this question, that is not good. If the answer is no overtime, that should be an obvious indicator that perhaps the need for more people could be met with overtime. If the group is working significant overtime already, it may support a request for increased headcount, but it also indicates a need to explore the factors driving the overtime (which, by itself, is a significant cost driver).

Determining Appropriate Headcount

How do we go about defining headcount in an objective, quantitative manner? Figure 2.1 at the beginning of this chapter shows a suggested approach.

The first step is to define the required tasks. This involves listing each step in the process. The organization will usually have this information available in its work instructions. If it doesn't, the people who do the work can describe what they do, or someone from the cost-reduction team or the work center can observe it and list the steps. Suppose, for example, a work center assembles mechanisms consisting of six parts (the base and five parts to be installed in the base). The tasks to assemble the mechanism include:

- Opening the work order (the document defining how to assemble the device) and reading it.
- Obtaining all six parts.
- Positioning Part 1 (the base) on the workbench.
- Installing Part 2 into the base.
- Installing Part 3 into the base.
- Installing Part 4 into the base.
- Installing Part 5 into the base.
- Lubricating all parts.
- Installing Part 6 (the cover) into the base.
- Testing the assembly's function.
- Completing the work order by signing and dating it.
- Moving the completed mechanism to a storage shelf.

After listing these steps, the next step is to assess how long each step should take. For new work, the assessment will be an estimate. If the work is already occurring, it might make sense to observe the operation and time it. If this cannot be done, however, estimating the required time is usually good enough.[1]

The next step is estimating the operator's efficiency when performing each task. The idea behind efficiency is to express, on a percentage basis, how quickly the operator will perform the task compared to the estimate developed earlier. We'll describe efficiency in more detail in the next chapter, but for now, it's enough to recognize tasks will generally take longer than the task estimates. That's because people need to go meetings, they need to go to the bathroom, they may work more slowly at times, different people will work at different rates, and so on. As a starting point, estimating efficiency at 80 percent for an initial cost estimate usually works well.

We next need to divide the estimated task times by the estimated efficiencies to obtain an efficiency-adjusted cost estimate. In other words, if an operator is 80 percent efficient at performing a 1-minute task, the efficiency-adjusted cost estimate is:

$$1 \text{ minute}/.8 = 1.25 \text{ minutes}$$

Note that we are expressing costs in terms of time. It's a simple matter to convert these to dollar amounts later based on the labor cost, but for now, we'll stay with using time as the way to express cost.

After computing the cost estimate for each task, we next sum the efficiency-adjusted costs obtained earlier. This gives the amount of time required for each assembly.

After obtaining the cost per assembly, we next need to divide the time required for each unit into the number of hours available during the workday. Typically, this is 8 hours. This calculation will give us the number of units an operator should be able to complete in a workday.

This approach provides a quantified estimate of what an operator can produce in a day, but we are interested in calculating how many operators are required. To get this, we need the work center's required daily output. The required daily output is usually dictated by the production schedule, which is in turn driven by the commitments the company makes to its customers. When we know how many units are required in a day, we divide this requirement by an operator's projected daily output. The resulting quotient is the required

---

1. Most people are hesitant to estimate the total amount of time it takes to complete a job. However, when jobs are broken down into their component steps as outlined here, most people feel comfortable estimating the time required to complete each step. Our experience indicates most people tend to underestimate the time required for each step (surprisingly, this even includes the operators doing the work), so caution is necessary.

headcount for this product in that work center. Table 2.1 shows how this is done.

The approach shows the required headcount for one product in one work center. Determining the required headcount for the entire operation requires repeating the analysis for each product that goes through each work center, and then summing the total for all work centers. It's a great approach for eliminating subjectivity in determining how many people are required.

Table 2.1 Headcount Determination Example

Task	Task Description	Time Estimate (minutes)	Efficiency Estimate (%)	Efficiency-Adjusted Time Estimate (minutes)
1	Obtain and read work order	4.00	80%	5.00
2	Obtain all six parts	4.00	80%	5.00
3	Position Base	2.50	80%	3.13
4	Install Part 2	4.00	80%	5.00
5	Install Part 3	5.00	80%	6.25
6	Install Part 4	2.50	80%	3.13
7	Install Part 5	6.00	80%	7.50
8	Lubricate parts	2.00	80%	2.50
9	Install Cover	3.50	80%	4.38
10	Test	2.50	80%	3.13
11	Complete work order	2.00	80%	2.50
12	Move mechanism to storage area	5.00	80%	6.25
	Total Time Per Assembly (minutes)			53.75
	Total Time Per Assembly (hours)			0.90
	Individual Worker Daily Time Available (hours)			8.00
	Projected Operator Daily Output			8.93
	Total Quantity Required Daily			68
	Estimated Required Headcount			7.61

Notes: Total Time Per Assembly (hours) is obtained by dividing Total Time Per Assembly (minutes) by 60 (there are 60 minutes in 1 hour). Projected Operator Daily Output is obtained by dividing Total Time Per Assembly (hours) into 8 hours (there are 8 hours in 1 day). Total Quantity Required Daily will typically be dictated by the production schedule (it is an input to this analysis). Estimated Required Headcount is obtained by dividing Total Quantity Required Daily by Projected Operator Daily Output.

At this point, after considering the explanation, you might be thinking that quantitatively determining headcount in this manner is a lot of work. It is. Consider the alternative, though. If you don't do it this way, your understanding of how many people the organization needs is at best a subjective guess. If you guess wrong, you either will have too few people and not get the work done in a timely manner, or you will have too many people and your costs will be too high. The cost of either of those situations can easily exceed the cost of doing the analysis described here, not to mention the effect on the shipping schedule, the organization's profit, the ability to win new work, and customer perceptions.

Overhead Headcount Determination

You may be wondering about headcount determination in overhead areas, such as quality assurance, purchasing, finance, sales, and so on. These departments usually do not have product-specific work instructions of the type described earlier that define in detail the steps involved in each job. This does not preclude using this approach to calculate appropriate staffing levels. The work content in the overhead functions can be defined as discrete tasks, time estimates can be developed for each of these tasks, the number of times the tasks have to be accomplished can be estimated, and these can be used to calculate headcount.

Managing Headcount

After the described analysis has been done for all work centers and the overhead areas, it is a good idea to prepare an x-y plot showing required versus actual headcount. This can be done for the entire facility for managing the plant at a macro level, and it can also be done at the work center level to allow supervisors to track their authorized headcounts.

Figure 2.2[2] shows a headcount projection for the entire facility. These kinds of charts are useful for keeping headcount from creeping up. They are also useful for showing if the organization is behind in adding staff when additional staff is needed. Not having enough people can drive up overtime and adversely affect the delivery schedule, as will be explained in a later chapter.

---

2. The template for this chart and Table 2.1 can be downloaded at www.ManufacturingTraining.com.

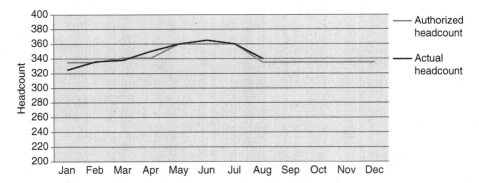

Figure 2.2 Authorized versus actual headcount. This graph shows at a glance how the organization's actual headcount compares to its authorized headcount.

The beauty of this approach in either situation (i.e., direct labor for manufacturing or overhead headcount) is that it forces an organization to think about the work content and express it in quantitative terms. People can debate how long a particular task might take and differences in individual performance, but they can't debate the identification of specific tasks and what the required outputs need to be. When debates about task times and efficiencies emerge, it is a good thing. It means the organization is thinking about how many people they need based on quantitative analysis.

The example described here and in Table 2.1 are rich in information, and this information suggests many opportunities. Consider the following points:

- The approach quantifies how many people are needed for a particular job in a particular work center. The projected headcount is driven by work content, required production rates, and projected efficiencies. It is not driven by gut feeling or a default response of "we need more people." It is quantitative. People can debate the finer points, but not the inherent logic.
- The approach can be used for developing headcount requirements for new work in a new work center, or it can be used to assess objectively the need for additional staff in existing work centers. It can also be used to assess if the headcount in a stable production environment is appropriate.
- The approach can be used to assess if overtime is necessary, or if existing overtime levels are reasonable.
- The approach can be used as a basis for developing labor costs in proposals for new business. When this occurs, the labor estimates for both direct and indirect costs will be much more accurate. As the proposal cost estimates are developed, they will reveal the task times that drive cost. With this visibility,

the organization can focus on reducing the cost of these tasks and become more competitive.

- The detail included in Table 2.1 suggests several cost-reduction opportunities. For example, can the organization accumulate completed jobs so that it only makes one move to the storage area, thereby reducing the time required? Can the organization pick up parts for more than one device at a time, similarly reducing the time required? Why does it take longer to install Part 5 than the other parts? Can the organization modify the process or the design to reduce this time?

- The analysis in the earlier example indicates the organization needs 7.61 people in the assembly work center. Should it have eight people or seven people? Can the work center staff at seven people and meet its required output through judicious use of overtime, or by sharing people with other work centers, or through process modifications, or with of efficiency improvements?

Who Should Do This Work

If the organization is large enough to have a manufacturing or industrial engineering group, these engineers are unquestionably the people who should prepare work instructions and time standards. If the organization does not have manufacturing or industrial engineers, the work center supervisors can prepare the time standards. The work center supervisors should be tasked with using this information for assessing appropriate headcount levels, and for other reasons that will be developed in the next several chapters.

Risks

When questioning headcount in existing organizations, people get nervous. Fear, rumors, and resistance are natural reactions. These are tough things to avoid, and there's no easy answer. The best ways we've found to control these reactions are transparency (be open about the effort) and speed. Assessing work content means that you have to understand and quantify the work, and you can't do that without people knowing what you are doing. We've found it's best to be open about it, and to let people know that in some cases you may find that more people are required, and in other cases, fewer will be required. Fear will exist, so it is best to not allow the assessment process to drag.

When you start examining what people do (their work content) they sometimes get even more nervous. In the direct labor area, this is easier to do because work instructions usually exist, and the product's design eliminates subjectivity about what it takes to make the product. In the overhead areas, it is more of a challenge. Many people are reluctant to describe what they do because of job security concerns. Many people in overhead functions cannot articulate their work content because they've never had to explain it (they know what to do, but they can't describe it). A few may be reluctant because they know they don't have much to do. Gentle questioning and a nonthreatening demeanor are always good approaches. Having the area supervisor explain the required tasks for each job is also a good approach. Sometimes this supports what you learn from the people doing the work. Sometimes it is better than having the person who does the work attempt to explain it if that person can't articulate what he or she does.

Another risk is getting it wrong. This can go either way. You might underestimate the work content, reduce staff, and not have enough people to do the work. Alternatively, you can overestimate the work content, hire too many people, and incur unnecessary cost. The best approach for mitigating this risk is to be as diligent as possible in identifying all tasks, assess the time required for each task as accurately as possible, and use reasonable efficiency estimates. This sounds daunting, but these estimates generally tend to be accurate. The headcount determinations tend to be much closer to actual needs when using the quantitative methods described here than they are if the managers and supervisors are guessing at what they need.

References

K. Zandin and H. Maynard, *Maynard's Industrial Engineering Handbook*, New York, McGraw-Hill, 2001.

G.W. Bohlander and S.A. Snell, *Managing Human Resources*, Florence, KY, South-Western College Pub, 2006.

3

Developing and Implementing Time Standards

The Bottom Line

Time standards define how long tasks should take. Time standards can be applied to both direct labor and overhead functions. Time standards should be developed using a blended approach of engineering estimates, historical performance, and actual observations. Time standards should be adjusted when they are found to be inaccurate. The organization should measure performance against the time standards and implement improvements where standards are not met. The time standards will reveal costly tasks, and the organization can use this information for targeting cost reductions. Time standards may be resisted for a variety of reasons, but the reasons are all invalid.

Key Questions

Do we have time standards?

How accurate are the time standards?

How are the time standards communicated to the people doing the work?

Do we compare actual performance to the time standards?

What do we do when we don't meet the time standards?

The Time Standards Development Road Map

A *time standard* defines how long a task should take. In the manufacturing world, time standards are stated for such things as machining a part, assembling a collection of parts, preparing parts for painting, painting parts, and so on. Time standards are prepared using techniques to be discussed later in this chapter. Many organizations do not have standards for the work they do, many organizations have inaccurate standards, and many organizations do not communicate standards information to the people doing the work. This chapter simplifies how to define, implement, and use accurate standards to reduce cost.

If no time standards exist, or if they exist but are not being used, the organization's efficiency is not being tracked, managed, or optimized. When this occurs, the organization is missing a significant cost-reduction opportunity. The reason for this is that there's a significant difference between telling someone to do something versus telling someone to do something <u>and</u> telling them how long the task should take. In the first case, the company has little control over labor costs (the job takes however long the operator takes to do it). In the second case, management is stating an expectation by defining how much time is budgeted for performing specific operations. When this is clearly communicated and measured, the actual time required tends to be much closer to the time standard. Where time standards are not used, actual performance varies widely.

Developing and using time standards is important for many reasons:

- Time standards clearly communicate management expectations to workers.
- The organization can accurately estimate labor costs when preparing bids.

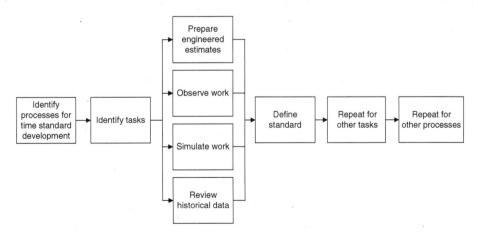

Figure 3.1 Recommended time standard development approach.

- The organization can assess its efficiency overall, at the work center level and at the operator level.
- The organization can determine if efficiency is improving or deteriorating.
- The organization can identify and target high-return efficiency improvement opportunities.

These considerations are critical in reducing manufacturing costs in any organization, especially those that are labor intensive.

Time Standard Development

Most organizations develop time standards using one or more of the following approaches:

- Engineered standards are based on examination and analysis of the drawings, equipment, and processes used for making components, subassemblies, and assemblies. This is similar to the example used in the preceding chapter for estimating the headcount required in a single workcenter for assembling a six-component mechanism.
- If the part or assembly has been produced in the past, the organization can use historical performance to identify the time required.
- If the part or assembly is in production, the organization can observe the process and base time standards on actual performance.
- If the part or assembly has never been produced, the organization can simulate the process and base time standards on the simulation.

Each approach has advantages and disadvantages. Engineered standards are strongly influenced by the estimators' competence. Standards based on historical or current performance are strongly influenced by past or current performance. If past or current performance has been inefficient, the inefficiency will be carried forward by the new time standard. Time standards based on simulations may not incorporate cost reductions due to learning curve or other efficiency improvements.

The best approach is to use a blend of the approaches and to temper the results with a healthy dose of common sense. These commonsense assessments include the following:

- If your organization is using an engineered standards approach, it should compare the engineered time standards to historical or current performance. If there's a difference, either the standards or the performance need to be adjusted to bring reality and the standards into agreement.

- If your organization uses historical data to develop time standards, it needs to assess if its performance is competitive. If it is not, both the actual performance and the standards need to be modified.
- Where the standards indicate high labor content, the organization should examine these operations to seek ways to reduce the labor. Much of the remainder of this book will be devoted to accomplishing this.

The bottom line to all of this will be a set of time standards for the organization's tasks, with each standard expressed in minutes or fractions of an hour for each operation. As mentioned earlier, this need not be restricted to manufacturing activities (although in organizations that have standards, it usually is); the concept of developing standards and measuring performance against the standards can also be applied to overhead functions.

If the organization already has time standards, the question to consider is:

Who prepared the time standards, and how good are they?

In some organizations, industrial or manufacturing engineers prepare the time standards. In others, manufacturing supervisors do. In yet others, other people in the organization do. In assessing time standard adequacy, we recommend considering the following:

- If the time standards are more than a few years old, it's likely they are outdated. Even if they still reflect current operations, older standards will not reflect expected learning curve improvements (we'll describe this in a later chapter).
- If engineers develop the time standards, how well do they know the process? Are the engineers frequently in the shop, or do they do all of their work from the office? Not being intimately familiar with the manufacturing operation is a serious issue in many organizations. Don't accept the standards as accurate simply because they were prepared by engineers.
- If the manufacturing supervisors develop the time standards, there may be a natural tendency to inflate standards so that they will be easy to meet.
- Do the people who do the work agree with the standards?

If standards exist but you suspect they are inaccurate, they should be corrected using the blend of technologies suggested earlier.

There's an argument that standards should never be adjusted once released, as this will make efficiency measurement and tracking meaningless. That's a hollow argument because standards are used in too many places (most notably, bidding new work[1]) to allow the continued existence of inaccurate standards. Standards should be adjusted as the operation improves, and management has a responsibility to improve the organization on a continuous basis.

If your organization does not have time standards, or the standards have fallen into disuse, you've uncovered a great opportunity. It's almost a certainty that cost-reduction opportunities exist and that implementing time standards will aid in acting on these opportunities. If either of these situations exists in your organization, implementing accurate time standards will focus the cost-reduction effort. A suggested approach for implementing accurate time standards includes the following:

- Target specific areas in which you wish to implement time standards initially, with others to follow. You won't be able to do the entire factory immediately.
- Select the people who will develop the time standards. If you don't have time standards, your organization may not have industrial or manufacturing engineers. You will have manufacturing supervisors, and they can develop the standards. You may also wish to consider retaining outside assistance for this task.[2]
- Define an approach for developing the time standards. We recommend a blend of the approaches outlined earlier (relying on past performance, preparing estimates of how long the tasks should take, and including cost-reduction goals).
- Develop and publish a plan for developing time standards, targeting processes in a sequential and logical manner.
- Meet to review progress on a weekly basis.

In this plan, you should recognize that developing time standards for the entire manufacturing facility is not an overnight activity; it is instead a long-term effort. It is achievable, though, if the standards development activity is approached in a reasonable manner.

---

1. Some organizations use the standards adjusted by the efficiency when bidding new work, which is another good approach. If the standard is known to be incorrect, however, it should be adjusted.

2. There are consultants who specialize in this area. This is also a great project to take to a local university with a manufacturing or industrial engineering department. Students can be employed at low cost, typically with the guidance and blessing of their professors, to develop time standards.

Using Time Standards to Reduce Cost

Time standards should be used to target cost reduction. Five steps are involved:

- Management or the cost-reduction team should identify where the costs are greatest, and focus on reducing the labor content in those areas.

- Management should clearly communicate the standards to the manufacturing organization. This lets everyone know what the expectation is. Approaches for doing so include specifying the time standard in the work instructions, posting the standards in the work center, and communicating the information verbally. In our experience, the latter two approaches work best (not everyone reads work instructions). Simply letting most people know how long they have to complete a task is effective in bringing the actual time into consonance with the time standard.

- The organization needs a tool for identifying the actual time taken to perform tasks. There are multiple approaches for doing so ranging from the sophisticated to the simple. Systems are available that use bar-coded entries on printed work orders, for example, to allow operators to scan in the start and completion of each task (and then the computer calculates the task duration so that it can be compared to the standards). The work center can simply count the number of jobs an operator completes during the day, divide that into the hours worked, and use the quotient to define the actual time required per job. The organization can require the operator to note the start and finish times for each job. In organizations that are initiating a time standards system, the latter two approaches are more likely to be used. Whether the organization opts for sophistication or simplicity is somewhat irrelevant; the important thing is to state the time standard, communicate it to the people doing the work, and then collect actual times to assess performance against standard.

- Once actual times have been collected, the organization can use these times to compare actual times to the standards to assess performance. This is used for evaluating efficiency, which is discussed in the next chapter.

- Where significant differences exist between time standards and the actual time required to do the job, the organization should investigate why. It may be that the time standard is inaccurate, it may be that the operator or department is inefficient, or it may be that a combination of both conditions exist.

In order for this approach to be meaningful, the organization should review overall performance against standards at a senior level on a monthly basis, with the data summarized along product and departmental lines. Individual manufacturing groups may wish to review the data more frequently, and they will probably want to do it at the operator level.

Who Should Do This Work

The team charged with developing the time standard should not be the same as the cost-reduction team. This is a specialized task and a separate team should address it. As mentioned earlier, the members should include industrial or manufacturing engineers (if the organization has them), the manufacturing supervisor from each area, and an independent review authority to make sure the time standards are reasonable (perhaps an engineer or someone from the finance department). Alternatively, the organization can seek outside support to assist in this area.

Risks

You will almost certainly encounter resistance when attempting to implement standards or drive the organization to improve the accuracy of and use existing standards. There are many potential sources and reasons:

- Resistance may come from manufacturing management. The manufacturing managers and supervisors may object based on the cost and effort involved in creating a standards program.
- If the organization has a bargaining unit (a union), it may object.
- Others may object based on fear of being measured or being held accountable for poor performance.
- Still others may object based on a fear that inaccurate standards will be used to punish adequate performance, or that differences in individual performance make the time standards meaningless.
- Sometimes salespeople object because standards objectively define costs and lead times, and using standards prohibits arbitrarily adjusting prices and delivery commitments.
- Similar objections will be raised when attempting to apply standards to the overhead functions. The principal arguments will be most likely that the tasks are too varied and actual performance is too difficult to measure.

Strategies and counterpoints to these arguments include the following:

- Although there are administrative costs associated with developing, implementing, and using time standards, consider the costs of not using standards. These costs include underestimating or overestimating jobs, allowing operators to take longer than necessary on manufacturing tasks, and having no meaningful tool for estimating new projects and measuring performance.

- The fear of being measured and being held accountable for one's performance or the performance of one's department should not be a factor in making a standards implementation decision. Performance measurement is inherent to running a business. People who object to this belong elsewhere.

- Concerns about using inaccurate standards for penalizing individuals or departments support the argument that standards should be accurate. These concerns don't support abandoning the effort.

- Individual performance will differ, but this argument is not a reason for avoiding standards. Differences are to be expected, but individual performance well below standard indicates an improvement opportunity. It may be that additional training, tooling improvements, reassignment to less complex tasks, or replacement is the solution.

- If the standards indicate that prices are too low or that quoted lead times are too short, this represents another improvement opportunity. The focus needs to be on finding ways to reduce task times and on understanding true costs. The organization should not ignore what the standards reveal.

- If overhead functions object to having standards applied to their activities, the reasons for the objection should be examined. As mentioned earlier, the frequently quoted reason is task variability and complexity. This has the feel of elitism to it. If the tasks vary, average task times can be used. The time required to process a purchase order, review a contract, or prepare an engineering drawing may vary from task to task, but over time, the averages in each category are relatively stable.

The challenge to a cost-reduction team wishing to implement a meaningful standards program is diplomatically responding to the previously discussed objections and persuading the people who object. Support from senior management is helpful.

One last risk remains and that is sustaining the effort. In many cases, organizations have standards but they have fallen into disuse. Monthly senior management efficiency reviews will help to prevent that.

References

F. Azadivar, *Design and Engineering of Production Systems,* Engineering Press, Inc., 1984.

J.P. Tanner, *Manufacturing Engineering*, New York, Marcel Dekker, Inc., 1991.

W.J. Stevenson, *Production/Operations Management*, Scarborough, Irwin, 1996.

4

Measuring and Using Efficiency

The Bottom Line

Efficiency is a measure of how well an organization is meeting the time standards. It is the ratio of the standard to actual performance. Poor efficiencies reveal significant cost-reduction opportunities. Organizations don't have to have sophisticated data collection or analysis techniques for collecting task times and calculating efficiencies; these tasks can be performed by work center supervisors. Efficiency trends should be tracked by operators and work centers to identify where efficiencies are improving, remaining flat, or deteriorating. Improvement efforts should be focused on areas showing no efficiency growth.

Key Questions

Do we measure our efficiency?

Can we identify areas of low efficiency and act on these to reduce cost?

Is a focus on efficiency adversely affecting product quality?

Is there a system that reports efficiencies on a regular basis, and do we use this information appropriately?

The Efficiency Measurement and Improvement Road Map

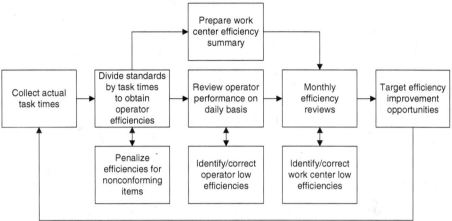

Figure 4.1 Recommended efficiency measurement and improvement approach.

What Efficiency Means

There are several definitions of efficiency. A common one is that efficiency equals output over input:

$$Efficiency = Output/Input$$

There are several problems with this definition, most notably including what constitutes input and what constitutes output. Some might consider input to be the number of hours worked, others might consider it to be how many people are employed, still others might consider input to be how much money is spent (there are other measures, too). Output is similarly ambiguous, at least from the perspective of how to reduce manufacturing costs. One measure of output might be sales, another might be amount of product shipped, another might be number of units produced, and so on. These are not bad measures (they all may have their place as benchmarks for assessing the organization's performance), but for our purposes, something more to the point is necessary.

In a manufacturing environment (and in particular, when seeking to reduce costs by measuring and improving efficiency) the preferred definition of efficiency is the ratio of the time standard to the actual time required to perform a task:

$$Efficiency = Time Standard/Actual Time$$

The reason this metric works in a manufacturing environment is that it allows supervisors to identify where efficiency is not what it should be. It allows focusing on inefficient areas.

As explained in the last chapter, there are several approaches for developing time standards, each with its unique advantages and disadvantages. As discussed, a blended approach using engineered time standards, historical performance, and simulations will provide the most accurate time standard. Collecting the actual times can be as simple as recording start and stop times for particular tasks or as sophisticated as implementing a bar-code reading system.

Evaluating Efficiency

Once the organization establishes standards and collects actual times, it can measure and evaluate efficiency. At the individual level, efficiency can be computed and used to assess operator performance. At the work center level, work center efficiencies can be compared to the time standards to assess work center and supervisor performance. Over time, efficiency trends can be analyzed to identify if efficiency is improving, deteriorating, or standing still.

Let's consider assessing individual operator performance first. Suppose we have a situation in which an operator is required to assemble a complex subassembly, and let's further suppose that the time standard for this task is 48 minutes. Over the course of a day, the operator completes six subassemblies, with actual times as indicated in Table 4.1.

Armed with these results, the question now becomes:

Is this operator performing efficiently?

Table 4.1 Operator Efficiency Calculations. The Efficiencies Shown for Each Subassembly Are the Time Standard Divided by the Actual Time

Subassembly	Time Standard (minutes)	Actual Time (minutes)	Efficiency
1	48	55	87.3%
2	48	58	82.8%
3	48	62	77.4%
4	48	71	67.6%
5	48	50	96.0%
6	48	52	92.3%
Totals	288	348	82.8%

One might initially conclude that the answer is no, because in every case the operator takes longer than the time standard to complete the task, which means the operator's efficiency is below 100 percent. That initial conclusion might be wrong, though, for a variety of reasons. All we know at this point is that the operator is not attaining 100 percent efficiency. Here are some of the possibilities:

- The time standard could be wrong. It might underestimate how long it takes to perform the task.
- Management may have unintentionally made the operator's life more difficult by locating tools, drawings, work instructions, materials, and other things required for the job far away from the operator's workstation. If the time required to obtain these items is not included in the original time standard, the operator will most likely not attain 100 percent efficiency (this is actually a subset of the topic listed rearlier; i.e., the time standard is wrong). In investigating these factors, a good supervisor will identify and take the steps necessary to eliminate these time wasters.
- The work area may have other detractors that induce delays, such as poor lighting, a frequent need for tool adjustments, and the like. A good supervisor will identify and address these factors.
- The operator may not know how to do the job. If additional training or work instruction improvements are needed, the supervisor should take steps to provide it.

The good news at this point is that we haven't gone very far in our quest for improving efficiency (and thereby reducing cost), and we already have a menu of things to investigate and potentially correct. They all came about by asking two simple questions:

How long should the task take?

and...

How long does the task take?

Similarly, we can assess efficiency at the work center level. Suppose the work center has five employees assembling the same product described earlier. Their performance on a particular day might be as shown in Table 4.2.

These results allow us to assess and target improvements at the work center level in the same manner as we did for an individual operator. Note that Johnson is attaining higher efficiencies than anyone else is. It might make sense,

Table 4.2 Work Center Efficiency. This Table shows the Efficiencies for All Operators

Employee	Sub-assembly	Time Standard (minutes)	Actual Time (minutes)	Efficiency
Smith	1	48	55	87.3%
	2	48	58	82.8%
	3	48	62	77.4%
	4	48	71	67.6%
	5	48	50	96.0%
	6	48	52	92.3%
	Totals	288	348	82.8%
Jones	1	48	67	71.6%
	2	48	66	72.7%
	3	48	74	64.9%
	4	48	75	64.0%
	5	48	80	60.0%
	6	48	68	70.6%
	Totals	288	430	67.0%
Nguyen	1	48	50	96.0%
	2	48	49	98.0%
	3	48	52	92.3%
	4	48	52	92.3%
	5	48	51	94.1%
	6	48	50	96.0%
	Totals	288	304	94.7%
Hernandez	1	48	54	88.9%
	2	48	52	92.3%
	3	48	60	80.0%
	4	48	58	82.8%
	5	48	49	98.0%
	6	48	54	88.9%
	Totals	288	327	88.1%
Johnson	1	48	46	104.3%
	2	48	47	102.1%
	3	48	48	100.0%
	4	48	46	104.3%
	5	48	47	102.1%
	6	48	50	96.0%
	Totals	288	284	101.4%
Total Work Center	Totals	1440	1693	85.1%

based on the results shown in Table 4.2, to closely study Johnson's work methods (assuming his work is of acceptable quality) to identify how he attains such high efficiencies. If Johnson has techniques or is using tools that shorten his actual times, this information could be used for modifying the process used by the other operators. Similarly, Jones' performance is below the others. Jones may need more training or other guidance to improve his efficiency.

The kinds of results shown here can also be summarized for different work centers, different steps in the process, and different products. Using these different perspectives will allow the organization to target and improve inefficient areas.

Evaluating Efficiency Trends

In addition to assessing efficiency at a specific point in time, the organization should collect efficiency data over time to assess whether efficiency is increasing, decreasing, or remaining flat (the last two situations should not be acceptable, as will be explained later). Suppose we track weekly work center efficiency for the month of July, as shown in Table 4.3.

Table 4.3 Efficiency Trend Data

| Work Center | Efficiency | | | | |
	3–Jul	10–Jul	17–Jul	24–Jul	31–Jul
Machining	82.0%	83.1%	81.0%	80.3%	79.4%
Paint preparation	75.5%	78.9%	81.2%	83.0%	82.4%
Paint	60.7%	59.6%	58.5%	58.4%	57.5%
Assembly	90.2%	91.3%	93.3%	94.5%	94.6%
Test	81.0%	81.5%	81.0%	81.3%	80.6%
Packaging	72.3%	74.0%	72.3%	73.3%	75.6%
Shipping	66.0%	66.3%	66.4%	66.6%	66.5%

We can then plot the data to assess trends as shown in Figure 4.2.[1] (For the purposes of this example, we are only taking a short peek at efficiency over a 1-month period. In actual practice, organizations track efficiency over longer periods of time.)

---

1. The graphs and tables included in this chapter can be downloaded at www.ManufacturingTraining.com.

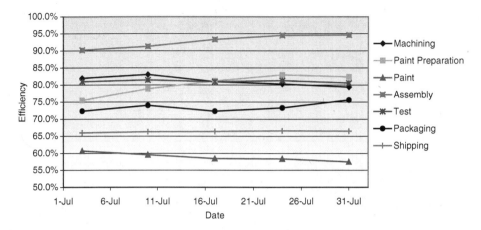

Figure 4.2 Work center efficiency trends during July. This chart shows that some work center efficiencies are improving, others are degrading, and others are essentially flat.

The chart shown in Figure 4.2 shows that the Assembly work center's efficiency is improving, as is the Paint Preparation work center's efficiency. This is a good thing (these work centers' costs per unit are coming down).

The Paint work center's efficiency is deteriorating, which is a bad thing (its costs are increasing). The supervisor should investigate why efficiency is decreasing in the Paint work center, and take appropriate corrective actions. Many things could be going on here. It might be that the Paint Preparation work center has stopped doing some tasks (which could explain the increase in their efficiency). This in turn means that the Paint work center has to do the tasks that the Paint Preparation work center has stopped doing (which could explain why Paint work center's efficiency is decreasing). On the other hand, it could be that the Paint Preparation area has actually improved while the Paint area has lost efficiency. More analysis is required to understand the reasons underlying these trends. The good news is that the trends are visible. The organization can use this information to focus the cost-reduction effort.

The other work centers, for the most part, appear to have essentially flat efficiency (it is neither increasing nor decreasing). This is not good, because it means these work centers are not improving. Over time, efficiency should increase and costs should come down. Management has a responsibility to make this happen. More investigation is required here to find out why the trend is flat.

One might examine the graph and conclude that the Assembly area's efficiency is substantially better than the other work centers. This may or may not be the case. It could be that the time standards for the Assembly work center are too liberal, and that those for the other areas are more aggressive. Alternatively, it could be that the standards in all areas are realistic and there actually are substantive differences in performance due to the processes, supervision, operator training, and so on. The good news is that using these evaluation methods allows for focused cost-reduction efforts.

In many cases, manufacturers worry less about the absolute efficiency values and more about the trends (as shown in Figure 4.2). Their reasoning is that inaccurate time standards will influence the absolute efficiency values but not the trend data, and the real interest is in efficiency improvement. In fact, some organizations do not change time standards once they have been established for this reason (they are more interested in trend data, and shifting time standards will artificially skew the trends). Although this logic has some merit, time standards should be as accurate as possible for the following reasons:

- Unrealistically tight time standards are demoralizing to operators, and frequently result in operators dismissing the time standard concept as a useful work measurement tool.
- Unrealistically liberal time standards can act to reduce efficiency, as operators may reduce output based on low management expectations.
- Inaccurate time standards (either too high or too low) result in inaccurate cost estimation, which can result in lost business or lost profits.

In our experience (and as discussed in the preceding chapter), manufacturers are best served by accurate time standards. Any minor perturbations to efficiency trend data resulting from accurizing time standards will most likely be minor, and in any event, identifiable.

Who Should Do This Work

A question that always emerges is who should do the work related to measuring efficiency. In larger organizations, the industrial or manufacturing engineers are normally assigned to standards development, and the efficiency calculations are automated. Usually the finance organization prepares and publishes the efficiency data. This makes sense for a few reasons. The first is that the finance organization is usually charged with collecting cost information, and the second is that having someone outside the manufacturing organization do the calculations assures objectivity.

As discussed previously, though, the absence of an automated system or a sophisticated data-collection capability should not preclude gathering and using efficiency data. In smaller organizations and in the absence of any system for data collection and reduction, the manufacturing supervisors can observe and record enough information to identify and act on efficiency-related cost-reduction opportunities. It's not that hard to do, and the better ones are probably already doing this.

In the overhead areas where such data are typically not collected, the supervisors can similarly be charged with evaluating efficiency. As in the earlier case, good supervisors should already be doing this.

Risks

Many of the same risks and risk management actions listed in the preceding chapter exist here.

Although few supervisors and managers will say it openly, those with poorly performing work centers will not want to publicize their inefficiencies. The way to address this risk is to focus on the reasons why we measure efficiency. The intent is not to punish people; the intent is to identify improvement opportunities.

Some may complain that measuring efficiency requires too much effort. This criticism can be blunted by pointing out that the organization will miss cost-reduction opportunities if efficiency is not measured, and that the cost of measuring efficiency is small compared to the likely return on this investment.

Some may complain that the people who create the standards do not understand the work. The way to manage this is to be diligent about correcting standards when errors are found. It's also a good idea to require supervisors to identify incorrect standards in their areas. Implementing this requirement lessens the criticism, because it places the responsibility for identifying incorrect standards on the supervisor. Complaining about incorrect standards then becomes an admission that incorrect standards are not being identified.

In the quest for improved efficiency, product quality may suffer. The risk here is that the organization focuses only on speed and ignores quality. A good way to manage this risk is to factor quality into the efficiency calculations. One approach is to not count rejected items in the efficiency calculations and to allocate the time spent on rejected items to other completed work. This significantly reduces efficiency. It forces those being measured to focus on both efficiency and quality.

Operators may resist efficiency measurement systems for the same reasons some supervisors and managers will (if they are inefficient, they won't want to broadcast it). A radical approach to managing this risk is to incentivize operators through bonuses as efficiency thresholds are met. This can work well for both efficiency and quality, especially if it is tied to quality as outlined earlier.

References

D. Korn, "Making Strides to Maximize Work Efficiency," *Modern Machine Shop*, September, 2006.

5

Assessing Machine Utilization

The Bottom Line

Utilization is the ratio of load to capacity expressed as a percentage. Machine and personnel utilization metrics show if the organization is effectively using its resources. Using this metric supports cost reduction by identifying areas of low utilization. Where low utilization occurs, management either should take steps to address it by shedding excess resources (if the workload is low) or by correcting low utilization causes (if work is available).

Key Questions

Do we measure utilization?

What are the sources of low personnel or machine utilization?

Is a focus on utilization adversely affecting efficiency?

The Utilization Measurement and Improvement Road Map

Utilization is a measure of how well an organization's labor and equipment are being used. It is the ratio of load (how much work the organization has) to capacity (how much work the machines or the workforce can do). The unit of measure for both load and capacity is typically hours, and utilization is usually

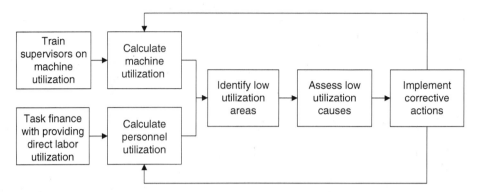

Figure 5.1 Recommended utilization approach.

expressed as a percentage. Utilization is usually evaluated on a weekly or monthly basis. The formula is:

$$\text{Utilization} = \text{Load}/\text{Capacity}$$

Utilization is important from a cost perspective for two reasons:

- If the utilization is low, the organization is not using its available capacity. This may occur because the organization has too little work. It may occur if the organization has too much equipment or too many people. If the utilization is low because there is too much equipment or too many people, the organization can reduce costs by reducing headcount or selling the excess equipment (assuming the company can't bring in more business to more fully utilize these resources). Low utilization can also occur because the work is there but it is not being assigned to machines and executed.
- If the organization's load is increasing such that load will exceed capacity, the organization needs to know this and act on it to prevent delivery delinquencies. The organization can purchase more equipment, hire more people, work overtime, add shifts, subcontract the work, decline to accept the additional work, or modify the process to increase the capacity.

Ideally, it would be great to walk into a factory and see all of the machines operating and all of the people in the factory making product. In this case, all resources are being fully utilized. On the other hand, it is disconcerting to see expensive machines sitting idle or people in the factory not working. The utilization concept helps to drive the factory and its workforce away from the second situation, and more toward the first one. In the process, costs decrease.

Machine Utilization

The machines in a factory under ideal conditions are available for as many hours as there are in a week. Each machine's capacity can be expressed as 7 days times 24 hours, or 168 hours per week. In reality, most companies work 40 hours per week on a single shift. Based on this, the capacity of each machine is 40 hours. This is further reduced by the need to maintain the machine and the set-up time (the time required to make the machine ready for the next job). Most organizations allow 15 percent for setup time and maintenance activities. If the machine operated 100 percent of the time it was available (recognizing that setup and maintenance detract from its availability), it would have a utilization of 85 percent, as shown here:

$$\text{Utilization} = (40 \text{ hours} \times .85)/40 \text{ hours} = 85\%$$

Based on this formula, 85 percent is often an upper bound for machine utilization. Management can do things to reduce setup time and perform preventive maintenance activities outside the normal 40-hour shift to raise this above 85 percent. Later chapters address these topics.

If the machine is not operated all of the time it is available, the utilization will be correspondingly reduced. For example, if the machine operates only 24 hours during the week, its utilization would be:

$$\text{Utilization} = (24 \text{ hours} \times .85)/40 \text{ hours} = 51\%$$

As seen in this calculation, utilization is easy to calculate (there's no rocket science here). Work center supervisors can note machine run durations and quickly perform the calculations indicated in this formula. Utilization can be considered from the perspectives of the machine, the operator assigned to it, the work center, or the entire factory. Utilization can also be trended to determine if it is increasing, decreasing, or remaining constant over time.

Assuming the work exists, measuring machine utilization is a good thing for the following reasons:

- It motivates supervisors to keep the machines running (i.e., making parts).
- Because of the machines running, it tends to reduce overtime.
- If a machine breaks down, it focuses attention on rapidly repairing it.
- It focuses supervision on using all resources.

Using the utilization metric has a downside, though. Seeking to increase machine utilization is only relevant if the work exists and it is performed efficiently. Simply relying on a utilization metric without considering workload and efficiency can result in artificial measures to bump up utilization (such as slowing machine feed rates, putting work on the machine when it is not needed, or making more parts than required). Such actions increase cost. They should be avoided.

Personnel Utilization

Management can also measure human resource utilization, which is a much better approach than only considering efficiency (a person could be extremely efficient when working, but work very little between jobs). In its simplest form, a quick gauge of utilization is simply counting how many people are making product versus how many are not. This is relatively easy to do. Just walk into the factory and count people, and then identify how many are engaged in making product.

In a more rigorous manner, supervision can collect the amount of time charged by a direct labor person to direct labor jobs, and divide that by the total number of hours available. Management should already be collecting this information so that costs can be assigned to projects or products. The same data can be used for assessing personnel utilization. For example, if an operator in the factory charges 32 hours to direct charge jobs, that worker's utilization is:

Utilization = 32 hours direct charge labor/40 hours available = 80%

In organizations that do not assign direct charge labor to projects or products, work center supervisors can still collect data and determine utilization.

The same caveats offered earlier for machine utilization are applicable here. If management focuses too heavily on utilization without considering efficiency, operators may tend to work more slowly or inflate the amount of time they work on a job. It may not just be operators who do this; sometimes their supervisors may direct them to do so.

Improving Utilization

The most obvious causes of low utilization are having too much capacity (too many people, too many machines, or both) or not enough work. If utilization is low, the organization needs to bring in more work or shed resources. Sometimes excess capacity may be a temporary condition (for example, if new work is coming in that will require the capacity later). In other cases, the organization may decide to accept low utilization temporarily to retain a trained workforce or for humanitarian reasons. In many cases, though, low utilization represents a cost-reduction opportunity.

Poor supervision can be the culprit when low utilization occurs in specific work centers. In this case, the supervisor may need counseling, tools, or guidance in how to assign work to operators or machines. In some cases, the supervisor may need to be replaced.

Poor process design resulting in upstream process bottlenecks can also induce low utilization in specific work centers. The upstream bottlenecks prevent work from getting to the downstream work centers, and downstream people and machines are underutilized. The organization can identify if this is occurring by examining utilization data for each work center. Sometimes this is obvious just by visiting the work centers. If successive work centers have high and low utilizations, it probably means the upstream work center is a bottleneck. Corrective actions may require process redesign or adding capacity (either people or machines) to the upstream work center. The fix may be as simple as moving people from the low utilization work center to the preceding high utilization work center (assuming the downstream work center folks have the required skills or they can be taught these skills).

Machine breakdowns often induce low utilization, especially if nonoperational machines remain dormant for extended periods. Machine breakdowns can result from:

- Inadequate preventive maintenance.
- Running the machine at too high a rate.
- Using the wrong machine for the job.
- Erroneous CNC (computer, numerically controlled) programs that cause machine crashes.[1]
- Continuing to use machines after their useful life.

Corrective actions for these problems include using the machines appropriately, assuring that the CNC programs are good, and a preventive maintenance program.

When machine breakdowns occur, an inability to quickly diagnose and fix the machine lowers utilization. This is a challenge in many organizations.[2] Management's stance should be that machines need to be fixed quickly; it should be

---

1. In our experience, this is the dominant cause of CNC machines not being available for use.

2. The issue here is to maintain an in-house machine repair capability or buy the services of outside repair specialists. Smaller companies generally use outside repair services, but this should not be a reason for accepting lengthy repair times. The fact is that both in-house and purchased services have significant disadvantages, including response time and cost. The best approach is a good preventive maintenance program to prevent breakdowns.

unacceptable to have a machine breakdown occur without rapid action to fix the machine.

Low utilization for specific individuals may point to operator-specific training or motivation issues, or poor supervision. Simply letting the supervisor know the low utilization situation is visible often corrects it.

Utilization as a Factor in Expansions

Utilization is a factor when a business is considering an expansion. If utilization is low, it doesn't make sense to hire more people or add more machines. Improving utilization is a much less expensive alternative. Whenever management considers adding more people or more equipment to increase capacity, the question should always be:

Have we done everything we could to improve utilization?

If the answer is no, the focus should be on utilization improvements prior to considering additional staff or equipment.

Who Should Do This Work

Surprisingly, utilization is frequently not tracked in both large and small organizations. Our advice is to have the work center supervisors do it for the machines, and have the finance organization do it for the people. Here's an approach that works well:

- Have the work center supervisors calculate utilization on each machine in their area. It's a straightforward matter to calculate what percent of the time a machine runs during the week. Two-decimal-place accuracy is not the intent. The intent is to get supervisors thinking about how to use all of their resources efficiently, and to push for rapid repairs when machines go down. Simply monitoring machine utilization usually improves it.

- The finance organization can quickly identify, by work center, the ratio of direct charge hours to total hours available. This information can usually be obtained in an automated manner. The concept here is to identify where direct charge factory labor is not being effectively utilized. With this visibility, management will know where excess labor capacity exists, and allow a focus on improving it.

Risks

Simply relying on a utilization metric without considering efficiency, quality, and other measures can increase costs. This might occur if machine rates are slowed, more parts than necessary are produced, or other artificial means are employed to bump up utilization. This can be prevented by training work center supervisors on how to use utilization along with other performance metrics, and by management evaluating utilization, efficiency, and quality simultaneously.

References

J.T. Womack, D.T. Jones, and D. Roos, *The Machine That Changed the World: The Story of Lean Production*, New York, Simon & Schuster, 1990.

W.J. Stevenson, William *Production/Operations Management*, Scarborough, Irwin, 1996.

6

Controlling Overtime

The Bottom Line

Uncontrolled overtime is a significant cost-reduction opportunity. Overtime is often abused because of poor planning, nonconforming material, greed, and other reasons. The organization can reduce overtime costs by establishing and meeting an overtime budget, identifying and resolving overtime drivers, moving overtime approval authority to higher management levels, and eliminating discretionary overtime.

Key Questions

Why do we need overtime?

How much overtime is appropriate?

Who authorizes overtime?

Do we have an overtime budget?

Should we pay overtime to our salaried workforce?

How can we reduce overtime?

The Overtime Reduction Road Map

Overtime can be a blessing to help meet short-term capacity shortfalls, or it can be a curse that relentlessly adds cost and inefficiency to an organization. A well-managed organization uses overtime to meet temporary capacity short-falls. These shortfalls can be seasonal (for example, working to meet a Christmas rush), or they can occur as the result of unusual events (a supplier fails to deliver needed components on time, internal process deficiencies impose an unusual rework requirement, or a client wants to pay an expedite fee to meet a required schedule acceleration).

One of the reasons managers tend to overuse overtime is because it is less expensive in the short run and easier than hiring additional staff. Overtime workers get a 50 percent premium for their efforts up to 10 hours of overtime per week, and a 100 percent premium for their efforts beyond 10 hours of overtime per week. That sounds high, but it is less than hiring another worker. Hiring an additional worker imposes a 100 percent premium plus a 40 to 50 percent benefits cost, plus the cost of the recruitment and interviewing effort.

This logic falls apart, however, if improvements to efficiency, utilization, and other factors are not considered in lieu of routine overtime, and it further deteriorates if the overtime continues for long periods. Long-term overtime degrades

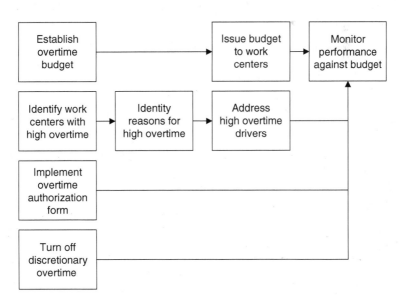

Figure 6.1 Recommended overtime management approach.

efficiency, which adds cost. In our experience, overtime lasting more than a few weeks is not efficient.

When used judiciously, overtime works well. The problem is that many organizations routinely work excessive overtime for months and sometimes years on end. People who routinely work overtime for long periods tend to tire and lose efficiency. Overtime can be used in a cost-effective manner. The problem is that many times it is not; where this is occurring, there are cost-reduction opportunities. These opportunities can be huge.

Overtime Abuse

People tend to misuse overtime for a number of reasons:

- Sometimes overtime occurs as a result of quality deficiencies. Overtime is needed to remake or rework nonconforming material.
- Sometimes people work overtime because they worked inefficiently during the day and they need to catch up.
- Sometimes overtime occurs because the supervisor planned poorly, and the work fell behind.
- Sometimes overtime occurs simply because people want the extra money. This becomes a particularly pronounced problem when people work overtime for extended periods. They become accustomed to the extra income and are reluctant to give it up.

Wherever these conditions exist, they can be eliminated or significantly curtailed. When this happens, costs decrease.

Establishing an Overtime Budget

Establishing an overtime budget is a critical first step in reducing overtime costs. An overtime budget should be established by asking the same questions asked for all other budgets:

- What are the anticipated costs?
- When will they occur?
- Where will they occur?

Answers to these questions should come from the work centers and be tempered by critical management review. There are no absolute rules for what constitutes an acceptable overtime level, and the required level will vary from work center to work center. That said, our general guidance is that overtime should not exceed 5 percent of the total hours worked.

Overtime is an expense, and it needs to be estimated, budgeted, and controlled. The size of the budget will vary based on the situation, but the mere presence of a budget helps everyone realize that overtime dollars are not infinite. At whatever level in the organization overtime approval authority exists, the budget should be considered when making the decision to authorize overtime.

Controlling Overtime

If operators or their first-level supervisors have the authority to decide to work overtime, that's a bad situation.

Our advice to manufacturing organizations is to always put the approval authority for overtime at a more senior level in the organization, at least one or two steps higher than the first-level supervisor. This alone makes it difficult to request and obtain overtime, and that is a better situation than simply allowing overtime whenever someone feels they would like it.

Whenever overtime is requested, it should not be a simple "Can I have overtime?" question followed by a yes or no answer. A good approach is to require anyone requesting overtime to fill out a request (see the example in Figure 6.2). This forces the supervisor who wants the overtime to quantify it and think about why it is needed, and hopefully, to find a way to get the work done without resorting to overtime.

An overtime request form makes the requestor quantify the number of hours required and identify the work that will be completed on overtime. It requires evaluating why the work could not be completed during normal work hours. It highlights efficiencies of the people who should have completed the work as well as the people who will do the work on overtime (there are few things more egregious than rewarding inefficient workers with more hours at a 50 percent overtime premium). The effect of doing all of this is not punitive, nor is it intended to bureaucratize the overtime approval process. The intent is to make it more difficult to get overtime and to prevent a cavalier approach to racking up overtime. It forces a focus on why overtime is needed and it identifies improvement opportunities.

Whenever overtime is used, the work center supervisor should check the next day to see if the work planned for completion on overtime was completed. Overtime work occurs after hours, and normal supervision will probably not be present. Knowing that output will be checked the next day helps to assure the company is getting what it paid for with its overtime dollars.

Overtime request	
Work center	
Supervisor	
Overtime hours requested	
Anticipated overtime cost (hours * labor rate)	
Work to be completed on overtime (number of units, specific operations, etc.)	
Operators who will work overtime	
Operator 1:	Efficiency:
Operator 2:	Efficiency:
Operator 3:	Efficiency:
Operator 4:	Efficiency:
Operator 5:	Efficiency:
Operators assigned to work during normal working hours	
Operator 1:	Efficiency:
Operator 2:	
Operator 3:	Efficiency:
Operator 4:	Efficiency:
Operator 5:	Efficiency:
Reasons work not completed during normal working hours:	
Alternatives to working overtime	
Work center supervisor signature and date	
Next level supervisor signature and date	

Figure 6.2 Overtime request form. The intent of using a form like this is to avoid unnecessary overtime, and to force work center supervisors and others to address overtime abuse.

Salaried Overtime

Most organizations do not pay salaried personnel overtime. The thinking is that people who are on salary are expected to work to get the job done, and that they are more self-directed. The idea is that salaried personnel do not have their pay docked if they need to take an occasional afternoon or morning off for personal reasons. It is expected that they will make up the time and meet their work objectives. The nature of a salaried position is that it is self-directed. Salaried employees do not need to account for every hour they work.[1]

Salaried overtime is difficult to manage. Organizations that pay salaried overtime find themselves in a dilemma. Most salaried professionals routinely work more than 40 hours each week (without being paid for it). They do this for a number of reasons, not the least of which is that salaried employees are usually paid much more than hourly workers. So, what happens when an organization pays salaried overtime? Does it need to pay for everything beyond 40 hours?

When an organization pays salaried overtime, costs increase significantly. In effect, the organization is paying for something it would probably get from the salaried workforce anyway. In organizations that pay overtime to the salaried workforce, our experience is that the salaried overtime is abused more than hourly overtime, probably because of less oversight and the inherently greater difficulty in measuring and assessing the salaried workforce's efficiency.

Here's a situation demonstrating what can be encountered in an organization that pays salaried overtime. In an aerospace company that designed and manufactured composite structures, salaried overtime in the engineering area was out of control. The organization made repeated efforts to limit the overtime with no success. Things reached a tipping point when one of the engineers commented to the plant manager that he was "saving work for the weekend in order to get overtime."

That rather startling admission brought the problem into focus, and management eliminated salaried overtime the next day. That weekend and for several following weekends, no salaried workers were in the plant. Amazingly, there was no decrease in output. The engineers still completed all of their work on schedule. After a while, salaried personnel started working unpaid overtime sporadically, but only when it was needed because of unusual circumstances.

---

1. There are exceptions, most notably on government contracts.

Our advice is this: Don't pay salaried overtime.[2] If salaried workers have to work significant overtime for extended periods, the work needs to be redesigned or the organization needs to hire more people.

Overtime Drivers

Several common overtime drivers exist. Most of these are obvious once explained; others are somewhat more abstract. The good news is that addressing these overtime drivers is relatively easy. The following paragraphs outline common problems, along with suggested actions to reduce the need for overtime.

- Poor planning often results in work center bottlenecks, with too much work arriving at specific work centers. The work centers cannot complete the work with their existing capacity. When this occurs sporadically, it can be managed with occasional overtime. If it occurs regularly, the organization should improve its production scheduling or the work center should add capacity.

- Poor quality results in overtime because the nonconforming material has to be reworked or remade. The corrective action here is straightforward. The organization needs to identify and eliminate the nonconformance sources.

- Supplier delinquencies can induce overtime. If suppliers do not deliver required materials on time, the process will not be able to proceed as planned. When the supplier materials arrive, overtime may be required to allow the organization to get back on plan. If this occurs infrequently, overtime can be used to address it. If it occurs regularly, the organization should work with the delinquent suppliers to eliminate their delinquencies or the organization should find new suppliers.

- Inefficiency or poor utilization can induce overtime. In some cases, the inefficiency or low utilization exists for technical or process reasons. In other cases, poor supervision induces it. In either situation, the organization should identify and eliminate the causes.

- Sometimes individuals create situations in which overtime is required because they want the overtime pay. Supervision needs to actively assess individual and work center efficiency and utilization to determine if this is occurring, and then take steps to eliminate it.

---

2. There are exceptions, particularly in cases where management directs the overtime and, as previously mentioned, on government contracts. The human resources department or a labor attorney can provide more specific guidance. Even in these situations, though, the overtime needs to be carefully controlled and executed in accordance with an overtime budget.

- The company may not understand its loads, capacities, and processes well enough to accurately define lead times (how long it takes to deliver the product after accepting the order), or the sales department may commit to deliveries below lead time in order to get the order. If the organization takes orders below lead time, this will almost certainly require overtime. This practice also induces delinquent deliveries. If this is occurring, the organization needs to either stop it or take steps to reduce lead time through efficiency improvements, utilization improvements, capacity increases, or process changes. In some cases, it may make sense to accept an order for delivery below standard lead time, but this should only be done in an informed manner with the required overtime identified prior to taking the order. When this occurs, most customers are willing to pay an expedite charge to cover the overtime.

Who Should Do This Work

Management should control overtime at all times, but this does not always occur. The cost-reduction team can serve as a catalyst to identify overtime as a significant cost-reduction opportunity. The finance department can point out areas where overtime is excessive. First- and second-level supervision in areas where overtime is occurring are key to reducing overtime cost; they must be onboard with the overtime reduction effort to make this happen. Senior management support for overtime reduction is also critical.

Risks

In any quest to reduce overtime, obstacles and resistance will emerge. These are the common ones, along with recommended actions to mitigate the risk.

- Handing out overtime may be viewed as a favor, and supervision may resist restrictions on their ability to do so. Senior management direction is essential for eliminating this risk.

- As mentioned earlier, many people adjust their lifestyles to living on overtime pay and will view any attempts to restrict overtime as a threat to their lifestyle. Morale may suffer. This should not deter management from eliminating unnecessary overtime, and the greed of individuals accustomed to padding their income cannot take precedence over the organization's profitability. Overtime is a cost driver, and controlling it is critical in the quest to reduce costs.

- As overtime is restricted, there is a risk that the work may not get done on time. The key to controlling this risk is good management. Understanding workload and capacity, taking steps to improve efficiency, and the other actions outlined in this book will allow management to intelligently determine when overtime is necessary.

- Overtime serves as a buffer for poor management practices, and supervision may perceive they won't be able to do their jobs as their ability to hand out overtime comes under tighter control. Establishing and managing to a budget helps to control this area. Most supervisors ultimately find that their fears are unfounded. When less overtime is available, the work in most cases still gets done.

- If restrictions on overtime interfere with the organization's ability to routinely deliver items below published lead times, the sales organization and customers may resist or complain about the overtime restrictions. Process, efficiency, and utilization improvements can alleviate this concern because they will shorten lead times. If these have already been implemented and overtime is truly necessary to meet a request for an expedited delivery, the organization can consider expedite fees. Most customers understand this and are willing to pay expedite fees if they are reasonable.

References

R.A. Hart, *The Economics of Overtime Working*, Cambridge, The Press Syndicate of the University of Cambridge, 2004.

H.R. Thomas, *Effects of Scheduled Overtime on Labor Productivity*, Austin, University of Texas at Austin, 1994.

7

Making Multiple Shifts Productive

The Bottom Line

When the workload on the first shift increases such that it no longer has adequate capacity, it may be necessary to start a second or even a third shift in selected work centers. Organizations should do so only after evaluating all other options for increasing first-shift capacity, as second and third shifts tend to be less efficient. Second- and third-shift efficiency, quality, and utilization should be closely managed. The need for multiple shifts should be revisited regularly to assure they do not outlive their usefulness. After implementing a second or third shift, the organization should continue to improve first-shift capacity where it is cost-effective to do so, with a view toward eliminating the need for multiple shifts.

Key Questions

Do we need a second or third shift?

Do the reasons that led to a second or third shift still exist?

How do we manage efficiency, quality, and utilization on the second or third shift?

Is there anything we can do on the first shift to eliminate the need for a second or third shift?

The Multiple-Shift Assessment and Management Road Map

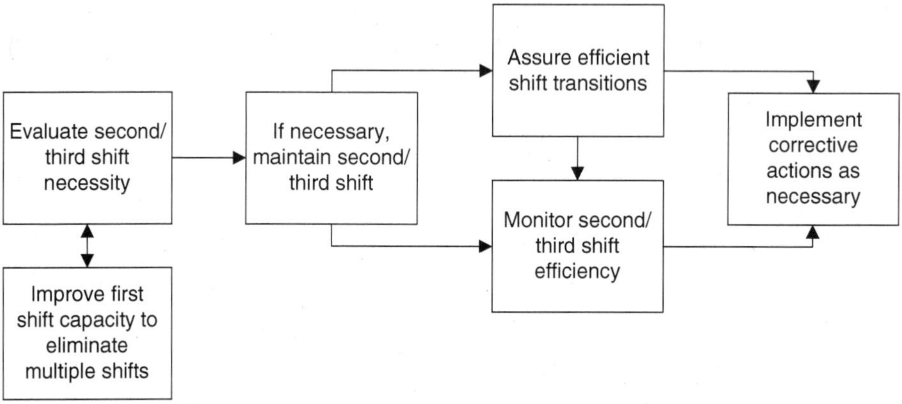

Figure 7.1 Recommended multiple-shift management approach.

Adding another shift can dramatically increase an organization's capacity. Before deciding to add another shift, however, the organization should quantitatively determine if doing so is justifiable. When second and third shifts are added, their efficiency must be monitored as closely as the first shift to keep costs down. This chapter explores criteria for making the decision to move to multiple shifts, and addresses how to reduce cost in a multiple-shift environment.

All manufacturing organizations have a first shift. The term *shift* refers to the weekly time period during which people are in the factory and work occurs. Usually this consists of five 8-hour days, with the shift starting in the morning and ending in the late afternoon (this is the first shift). Some organizations run a second shift (another 8-hour shift, starting immediately after the first shift ends), and a few even run a third shift (with its 8-hour cycle immediately following the second shift).

The Multiple-Shift Decision

Sometimes, the first shift does not have enough capacity to get all the work done. As the workload increases beyond the first shift's capacity, the options are:

- Improve the efficiency, utilization, and processes such that the first shift's capacity increases. This is the ideal approach. The rest of this chapter assumes this has already occurred. You should not make this assumption when considering adding additional shifts unless you know it has happened and no capacity-increasing, first-shift improvement opportunities exist.

- Use first-shift overtime.
- Add additional equipment and hire more people on the first shift. Sometimes this can be done with existing facilities; if this is not the case then additional facilities are required.
- Add a second shift (and perhaps even a third shift), using the equipment and facilities used by the first shift.
- Off-load the work to one or more subcontractors.
- Extend the lead times on new work. This allows using the existing capacity, but pushes deliveries further into the future.
- Decline to accept the additional work. The assumption here is that the organization has already decided to accept the additional work.

All of these options should consider the nature of the required capacity increase. If it is a temporary situation, the organization is probably best served by using overtime. If the workload increase is not temporary and it cannot be met with overtime or efficiency improvements on the first shift, it probably makes sense to add people and equipment on the first shift if facility constraints allow doing so.

Extending the lead time is another option. This allows using the existing capacity, but it moves the delivery date further into the future. In some cases, customers can accept delayed deliveries. For competitive reasons, however, this is often not an attractive option.

If the workload growth and required capacity increases are such that the work cannot be accomplished on the first shift (even with efficiency improvements, overtime, more people and equipment, or longer lead times), the organization needs to either add shifts or subcontract the work to other organizations. Subcontracting the work to another organization involves make-or-buy considerations, which will be discussed in another chapter. The decision to add another shift is our focus here. Before making the decision to add another shift, we recommend objectively evaluating the following questions:

Is the first shift working efficiently and is all equipment being utilized on the first shift?

We previously discussed approaches for making this assessment. If the first shift is not efficient, or if the equipment is not being appropriately utilized, it makes little sense to layer these inefficiencies with another shift. Intellectual laziness may make adding another shift seem attractive, because it is easier to add another shift than it is to address low efficiency or poor utilization on the first shift. This would be a costly decision, however, if the required capacity increases can be met with improvements on the first shift.

What is our absenteeism rate?

This is an extension of the efficiency and utilization issues discussed earlier, but it is important enough to merit individual focus. If the organization does not know its absenteeism rate, that is not a good sign. If absenteeism is high, this is also not good. In well-run organizations, absenteeism is low (typically well under 3 percent or so). If the absenteeism rate is higher than this, the absenteeism issue should be addressed before adding another shift. This is actually not a very difficult problem to solve. In many cases, simply letting it be known that absenteeism is being tracked reduces it. That won't completely eliminate the problem, and in some cases, disciplinary action may be required for habitual absenteeism. Terminating employees with high absenteeism rates lowers overall absenteeism for two reasons – the principal offenders are gone, and others who might have attendance issues realize that high absenteeism will result in job loss.

What is the cost of adding overtime on the first shift, and how does this compare to the cost of adding a second or third shift?

This question is not as straightforward as it might seem. Extended overtime results in reduced efficiency for all of the reasons described in an earlier chapter. To estimate the true cost of overtime, anticipated efficiency reductions should be included in the calculation.

Similarly, determining the cost of adding a second shift is also not as straightforward as it might seem. There are numerous costs associated with adding a second or third shift:

- The organization will incur recruitment costs, including the cost of finding, interviewing, and hiring people.
- Not all second-shift new hires will work out, and the process will have to be repeated for some of the second-shift positions.
- There will be a learning period associated with a new shift (during which efficiency will be low).
- Even after the second shift has had time to learn its trade, efficiency will probably still be lower than on the first shift.
- A second shift will require at least as much supervision as the first shift (anyone who thinks this is not the case is either wrong or they have too much supervision on the first shift).
- Utilities expenses will increase, because more energy will be required to power the equipment, lights, and other power consumers on the second shift.

- Second- and third-shift employees will have more health problems. Numerous studies show that a variety of stress-related illnesses occur more frequently with second- and third-shift employees, increasing absenteeism, and other health-related costs.

All of these factors must be quantified and considered when making the decision to add a second or third shift.

Second/Third-Shift Cost-Reduction Opportunities

Let's assume that after completing the analysis, the organization decides to implement a second or perhaps even a third shift, or that you find yourself in an organization already running a second or third shift. Given that this is the case, multiple cost-reduction opportunities exist.

- Transitions between shifts often result in poor continuity from one shift to the next. The organization should have an approach for defining how work is to continue from one shift to the next. This is particularly important if the nature of the product requires that the work spans shifts (this would occur, for example, on complex items that take longer than a single shift to complete). It might make sense to budget a half-hour of overtime for the first-shift lead person to oversee the shift transition. The overtime will add cost, but it will probably eliminate inefficiencies during the transition such that the overall effect is lower cost.

- Second and third shifts will almost certainly have lower efficiencies. Most people are, by nature, not nocturnal. Working until midnight (or starting the workday at midnight) runs counter to our natural circadian rhythms, and the efficiencies attained on a second or third shift will be somewhat lower than those attained on the first shift. That being said, it still makes sense to track efficiencies on all shifts. A differential of a few efficiency percentage points may be acceptable; a 50 percent decrease is not. Implementing a second or third shift without assessing its efficiency would be a costly mistake. Where significant efficiency shortfalls are uncovered, action is required to resolve them.

- Sometimes the conditions that led to implementing multiple shifts change, and the increased workload that necessitated a second or third shift no longer exists. Second and third shifts tend to take on a life of their own and become institutionalized. If multiple shifts exist, the organization should regularly examine the need for the additional shifts. If the need has disappeared (and the workload falloff is not temporary), the additional shifts should be eliminated. This sounds obvious, but in our experience, second and third shifts have a tendency to live beyond when they are needed.

- When second and third shifts are implemented, many of the support and other overhead functions that are available on the first shift will not be available on the later shifts. These might include calibration, repair, human resources, and other functions. If the absence of these functions on a second or third shift can impede efficiency or throughput, these issues should be identified and resolved before they have a cost impact.

- Organizations implement second and third shifts to eliminate the need for excessive overtime on the first shift. If first-shift overtime continues unabated after adding a second or third shift, something is wrong. The reasons for the continuing overtime on the first shift need to be resolved.

Who Should Do This Work

The manufacturing organization (assuming a second or third shift is required for production reasons) needs to take a lead role in assessing the cost, efficiency, and supervision issues outlined in this chapter. The finance organization should support the cost analyses required when making the multiple-shift decision. Support organizations (for example, the quality assurance group) should provide inputs to the multiple-shift decision, and should monitor the efficiency of people in their groups assigned to a second or third shift.

Risks

The biggest risk in running a second or third shift is in failing to address the efficiency shortfalls that frequently occur in these late-night work environments. As mentioned earlier, efficiency needs to be monitored on these shifts just as it is on the first shift.

The organization should recognize that some of the services available on the first shift may also be needed on the second shift. This can include inspection support, maintenance, and emergency services.

References

F. Azadivar, *Design and Engineering of Production Systems*, Engineering Press, Inc., 1994.

8

Finding and Reducing Lost Time

The Bottom Line

Lost time is all the time spent by direct labor personnel not making product. Lost time is usually not the result of disciplinary shortfalls; it is instead the result of forcing direct-labor people to spend time on activities other than making product. Delay ratio analysis is a technique that separates lost time from productive time. It identifies lost time and allows focusing improvement activities on reducing lost time. The delay ratio analysis is also useful for assessing office and overhead functions. Small lost time improvements can result in huge cost reductions.

Key Questions

How much of our time is spent actually making product?

What are the tasks that take time away from making product, and what can we do to eliminate or reduce this time?

The Lost Time Reduction Road Map

Most organizations are shocked when they learn that only about 15 to 20 percent of their total direct-labor work time is spent on actually making the

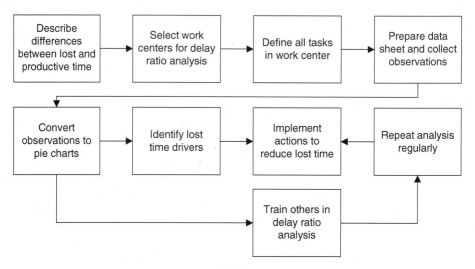

Figure 8.1 Recommended delay ratio analysis approach.

product. The remaining 80 to 85 percent of their time is lost time. Lost time is time spent on nonproductive tasks, such as picking up tools, gathering parts, talking, examining drawings, or simply not being present in the work center. This chapter presents a simple analysis technique called delay ratio analysis for identifying and quantifying the sources of lost time, and it provides recommendations for converting the nonproductive lost time into productive work time. Organizations frequently find that they can triple or quadruple their capacity and output by focusing on lost time reduction.

Seeing Lost Time

Lost time is any time spent by the manufacturing workforce not making product. That's a simple definition and it seems obvious, but most people have not conditioned themselves to seeing lost time.

Let's consider an example. Suppose you walk into the factory and see eight people, all of whom are busy in a work center that has ten employees. Of the eight:

- Two employees are running milling machines and cutting chips.
- One employee is reviewing a drawing and discussing it with a supervisor.
- Two employees are setting up lathes.
- One employee is stacking parts he recently completed on another mill.

- One employee is sweeping chips around a drill press.
- One employee is returning from the tool crib with the tools needed for the next job.

Most people would view this situation and conclude that life is good. Everyone is busy, the factory is productive, and no one is wasting any time.

They would be wrong.

As the situation is described, only two people are making product. The other eight (the one discussing a drawing, the two setting up machines, the one stacking parts, the one sweeping up, the one returning with tools, and the two who are not present) all represent lost time. They are not making product. It is this ability to differentiate between people actually making product versus those who are not that is critical to identifying and eliminating lost time. When we can make this differentiation and do something about it, costs decrease and output increases dramatically.

Some might argue that with the potential exception of the two people who are not present, all of the people listed in our example are performing necessary tasks and therefore their activities should not be considered lost time. The concept of lost time (and perhaps even the term) conveys the sense that someone is doing something wrong, perhaps even wasting time when they should be working. Certainly, people who waste time by socializing or engaging in other non-work-related tasks represent lost time. However, we don't want to limit the lost time concept to these more egregious circumstances. If someone is not making product, regardless of whatever else he or she is doing, his or her time for making product is lost. This time is what we wish to identify and reduce.

Delay Ratio Analysis

The delay ratio analysis is a concept that is extremely useful for identifying where opportunities exist to eliminate lost time. Here's how it works:

1. All activities in a work center are identified.

2. The activities are listed in a table, with the different activities listed on the left side of the table.

3. At random times during the day, the person preparing the delay ratio analysis notes the time at the top of each column in the table, views the work center, and lists the number of people doing each activity.

4. Steps 1–3 are repeated several times during the day, with the time again noted at the top of each column and the number of people engaged in each activity recorded in that column.

5. At the end of the day, the person assembling the analysis sums the total number of observations in each activity category.

6. The summed observations are presented in a pie chart showing the percentage of observations in each category.

7. The categories are divided into two groups: Productive Work and Lost Time.

8. A pie chart is similarly constructed showing the percentage of productive work versus lost time.

Let's consider a delay ratio analysis prepared for a paint shop. After visiting the work center, observing the operation, and speaking with the supervisor, you learn that the activities occurring in the paint shop include sanding, painting, moving material, loading the sprayers, cleaning the sprayers, reading the work orders associated with different jobs, adding written data to the work orders, obtaining sprayers from the tool crib, obtaining paint from the stock room, and cleaning the paint room air filters. To these activities, you add walking, talking, standing around (i.e., being present but not working), and not present. The work center supervisor explains that the work center includes fourteen employees (not counting the supervisor), and all are present.

Based on this, you can prepare a table as described earlier showing:

- The activities on the left side of the table.
- The times you make observations on the top of the table.
- The number of observations in each category each time you make a set of observations.
- The total number of observations in each category on the right side of the table.

Let's assume that over the course of a day[1] you observe the work center and record the information in Figure 8.2 as shown.[2]

This is good information, but it's going to get even better when it appears as a pie chart. The Microsoft Excel spreadsheet program makes doing this simple. Figure 8.3 shows the resultant pie chart.

---

1. Doing this does not take an entire day. Each set of observations takes only a minute or two. Most analysts take the observations when they take a break from their other duties.

2. The charts and tables included in this chapter can be downloaded at www.ManufacturingTraining.com.

Activity	Time											Total observations
	8:47	9:22	9:58	10:21	10:50	11:30	13:00	13:45	14:21	14:51	15:35	
Sanding	2	1	1		1	2	1	2	1	2	1	14
Painting	2	2	2	3	1	1	2	2	2	2	2	21
Moving material	1	2	2	3	1	1	1	2	3	1	2	19
Loading sprayers	1		1	1		2	1	1	1		1	9
Cleaning sprayers		2	1	1		1			1			6
Reading work order	1				1		1	1		2		6
Filling out work order		1	1			1	1	1	1		2	8
Obtaining sprayer	1		1		1		1					4
Obtaining paint	1	1			1	2						5
Cleaning air filters				1	1		1	1	1	2	3	10
Walking	2	3	2	1			2					10
Talking	2	2	3		3				2			12
Standing around				2	1		1					4
Not present	1	0	0	2	3	4	2	4	2	5	3	26
Total head count	14	14	14	14	14	14	14	14	14	14	14	154

Figure 8.2 Paint shop delay ratio analysis raw data.

The pie chart is interesting, because it shows several nonproductive activities that are relatively high:

- Not being present in the work center accounts for 17 percent of the observations.
- Talking accounts for 8 percent of the observations.
- Walking accounts for 6 percent of the observations.

These three categories are 31 percent of the observations, which implies that there's room for improvement. We're not done yet, though. Recall our earlier observation that any activity that is not producing product is lost time. Based on this definition, let's consider sanding and painting to be productive work, and everything else to be lost time. Figure 8.4 shows that pie chart.

The Figure 8.4 pie chart is extremely revealing. It points out that in a work center designed to paint product, only about 23 percent of the effort is actually being used to do so. The other 77 percent of the time is lost to obviously nonproductive tasks (standing around, not being present, talking, etc.)

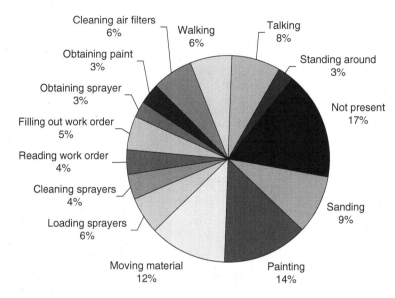

Figure 8.3 Paint shop delay ratio analysis pie chart. This chart shows all activities occurring in the paint shop based on observations occurring in one day.

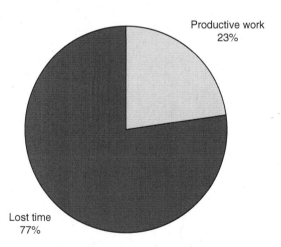

Figure 8.4 Delay ratio analysis pie chart. This pie chart shows the work segregated by productive work versus lost time.

and other required but nonproductive support tasks (cleaning filters, obtaining materials, etc.). What this shows is that substantial improvement opportunity exists.

This paint shop illustration is not just an example contrived to illustrate the concept. In our experience, the first time a delay ratio analysis is prepared for nearly any work center, it typically shows that only 15 to 20 percent of the people are actually making product. In many cases, the time spent actually making product is even less than 15 percent. That's a startling but not uncommon situation.

So, what can we do with this information? The good news is that small reductions in lost time typically result in huge productivity gains, and that means correspondingly huge cost reductions. To realize these, we need to consider and interpret what the lost time data reveal to us, and then think about how to act on these revelations. Here are a few paint shop interpretations and associated improvement actions:

- The Figure 8.3 pie chart indicates several instances of people walking, talking, and standing around. There were a total of ten, twelve, and four observations in these areas (a total of twenty-six observations). It should be relatively easy to make a dent in this lost time area through improved supervision and by relocating the things people need closer to where they are needed. Let's assume these actions cut this lost time by about one-third. If that happens, we would only have seventeen observations (instead of twenty-six) in the next delay ratio analysis for this work center. There's still room for improvement, but we'll show modest improvements here to make a point.

- The Figure 8.3 pie chart indicates numerous instances of people not being in the work center. This may be due to people socializing, or it may be due to people having to leave the work center for materials or other work reasons. If it's a discipline issue, it should be relatively easy to address. If the absences are work related, examining the reasons for those absences and correcting them should be straightforward. Corrective actions might include such things as locating required materials in the work center, keeping drawings in the work center, and so on. Good supervision and corrective actions of the nature described earlier can eliminate the reasons workers need to leave the work center. The initial analysis showed twenty-six instances of workers not being in the work center; let's suppose we can drop this by one-third as well. If we do this, we will have only seventeen observations in this category in the next delay ratio analysis.

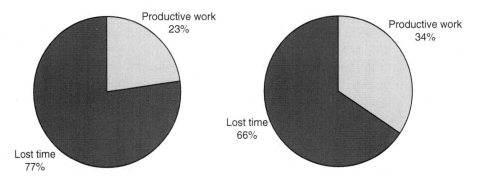

Figure 8.5 Delay ratio results before and after taking corrective actions. Note that productive work increased from 23 percent of the total to 34 percent of the total. This is nearly a 50 percent improvement in productive work.

Let's assume that one-third of the staff who were engaged in walking, talking, standing around, or not being in the work center can now be engaged in productive actions (sanding or painting). Figure 8.5 shows the original delay ratio pie chart, and then another from a delay ratio analysis prepared after taking the specified actions.

The productive work portion of the work has increased from 23 to 34 percent. One might be tempted to see this as only an 11 percent improvement (which is significant), but the improvement is actually much more. The increase in productive work is the 11 percent gain over the original 23 percent, or nearly a 50 percent improvement! One of the great things about delay ratio analysis (and the improvements it leads to) is that small decreases in lost time result in huge increases in productive work.

The delay ratio analysis approach is easy to use, and the payback is huge. It is a great tool for reducing manufacturing costs.

Who Should Do This Work

Anyone who ever walks through a factory does delay ratio analysis subconsciously just by assessing how many people look busy versus how many are standing around. The difference here is that looking busy is not the criteria. Making product is.

The delay ratio technique is simple, and just about anyone can do it. It probably makes sense for someone on the cost-reduction team to do the first few delay ratio analyses, and then train others to do it.

The interesting thing about this technique is that it is as applicable to office and other overhead functions as it is to manufacturing work centers.

Risks

Delay ratio analysis is easy to use and it often provides startling results. The only risks associated with the technique are that some people are threatened by it, some people may claim the results are not meaningful, and others may reject the distinction between the productive work versus nonproductive work categories.

People can feel threatened by the delay ratio analysis technique because the results indicate such a shockingly low percentage of the effort is going into productive work. Others may view it simply as a tool to find out who is goofing off. The best approach for reducing these reactions is to explain the difference between lost time and goofing off. The assumption here is that most people are working and most people want to do a good job. The challenge is identifying where their hard work is not going into making product. When presented in this light, most people accept the results. In fact, many are downright amazed at what the technique reveals and how it guides improvement activities. Successful organizations using this technique can see their productive work portions increase from 15 percent in the first analysis to perhaps 80 percent after a few weeks of process redesign, with a corresponding *533 percent*[3] increase in productive work!

Sometimes objections emerge because of the point sampling nature of the observations. If this is a concern, it is a simple matter to repeat the analysis at different times. Surprisingly, until the work center incorporates changes to address the lost time, the delay ratio analysis findings are usually very consistent.

Another criticism is that many tasks identified as nonproductive work are required to accomplish the job. This is often true, but it does not alter the fact that reducing the percentage of time spent doing these things translates directly into time actually making product. There are a number of things that can and should be done to eliminate or reduce these necessary activities that contribute to lost time. These actions are covered in the rest of this book.

References

E. Goldratt, and J. Cox, *The Goal*, Great Barrington, North River Press, 1992.

A. Mackenzie, and P. Nickerson, *The Time Trap: The Classic Book on Time Management*, AMACOM, 2009.

---

3. Based on the productive work portion of the total increasing by the ratio of 80 percent/ 15 percent.

9

Using the Learning Curve to Drive Costs Down

The Bottom Line

The organization can use learning curve theory to drive costs down by predicting future cost reductions, monitoring actual costs against the prediction, and taking actions to address any variances. Learning curve theory predicts that as cumulative production quantities double, the cumulative average cost will decrease at a constant rate. The theory is well proven and applies to all industries. The rate at which costs decrease is expressed as the learning curve percentage. Lower percentages mean sharper cost reductions.

Key Questions

Are we using learning curve theory?

What is our learning curve, and how did we arrive at this number?

How do our current costs compare to the learning curve prediction, and if there is a variance, what are we doing about it?

The Learning Curve Road Map

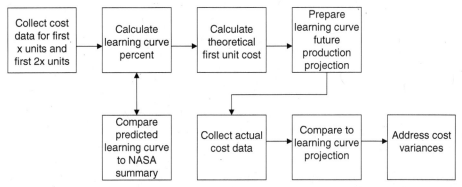

Figure 9.1 Recommended learning curve approach.

Learning curve theory is a powerful tool for driving costs down. The theory maintains that learning always occurs, and as it does, costs should decrease at a predictable rate. If costs are not coming down (or they are not coming down quickly enough), costs are too high.

Those who don't understand learning curve theory or are intimidated by its mathematics may be tempted to dismiss it. Others may claim that learning curve theory doesn't apply to their industry, because their products or processes are "special." These folks are missing an important opportunity. The existence of the learning curve is irrefutable. The rate at which the learning curve causes costs to come down may vary based on any of several factors, but the learning curve always exists.

The Learning Curve

Learning curves are expressed in terms of a percentage (e.g., a 90 percent learning curve), and this percentage expresses how sharply costs decrease. If we say that a product has a 90 percent learning curve, what this means is that the cumulative average cost decreases by 10 percent each time we double the cumulative production quantity. Based on this definition, as the learning curve percentage decreases the cost decreases more sharply. An 80 percent learning curve means costs come down more sharply than a 90 percent learning curve.

We'll describe how to predict and calculate learning curves based on the product type and labor content, but for now, it is important to recognize that the learning curve is a powerful tool for predicting and measuring the effect of production quantities on cost reduction. If the organization has not predicted its learning curve, or if it applies the same learning curve to all products, it isn't using learning curve theory well (we'll discuss how to develop a learning curve and the factors that influence it later in this chapter).

The mathematics behind the learning curve theory are a bit more complex than those discussed thus far in this book, but they are not so difficult as to be unusable. Before delving into the mathematics, let's first make sure we understand what they describe. In its simplest terms, the learning curve means that as quantity increases, the cost per item decreases, and it does so at a predictable and consistent rate regardless of how many items are produced. Stated differently, costs will continue to decrease if you keep making the product for as long as you continue to make it. Several points are relevant:

- In the context of this discussion, cost refers to labor hours, although it can be converted to dollars.
- The quantities we discuss when using learning curve theory are cumulative quantities (in other words, the total amount produced from the start of production).
- The costs we discuss when using learning curve theory are average costs for a group of parts, as we'll see later.
- The cost decreases in accordance with an exponential function. We'll explain what that means shortly.
- For each doubling of the cumulative production quantity, average cost is reduced by a constant percent. This percentage is the number we use when we talk about the learning curve. For example, if we say a product has a 90 percent learning curve, it means that each time we double the cumulative production quantity, the average cost of the entire quantity will be 90 percent of the average cost of the first half. As explained earlier, lower percentage learning curves mean that costs are reduced more sharply. We describe this by saying the learning curve is steeper. This is somewhat counterintuitive, as higher percentages usually imply greater steepness when describing other things.
- When we describe the percentage, what it means is 100 percent minus the rate at which the cost decreases. If cost decreases 20 percent each time the cumulative production quantity doubles, an 80 percent learning curve results ($100\% - 20\% = 80\%$).

With this in mind, here's how we describe the learning curve mathematically:

$$CC = (C_1)(N)^{-(p)}$$

where:

CC = Cumulative average cost
C_1 = Cost of the first unit
N = Total number of units produced
p = An exponent to which the number 2 is raised to give the learning curve percentage

Learning Curve versus Work Content

As a first step, it is a good idea to determine the learning curve for your product. If your company has built similar products in the past, it is likely that those products' learning curves will be similar to the product for which you wish to establish a learning curve. Either you can use published sources for the type of work your company does, or you can calculate your own learning curve percentage based on historical results.

The National Aeronautics and Space Administration (NASA) publishes recommended learning curve percentages based on work mix (as shown in Table 9.1) and based on industry (as shown in Table 9.2).

Note that in Table 9.1 as assembly content (i.e., hand labor) decreases, the learning becomes less steep. This is also apparent in the data that NASA publishes for recommended learning curves by industry, which Table 9.2 shows. The reason for this is that labor is where the greatest learning occurs (machines generally don't get faster at what they do, but people do).

The data in Table 9.2 are a good place to start, and the data are useful if your organization has no production history for the product to which you wish to apply learning curve theory.

In most cases, though, organizations do not introduce completely new products. Most products are evolutions of existing products. Given that, it's relatively easy to estimate a learning curve specific to your products.

Table 9.1 NASA Learning Curves Based on Work Mix

Work Content	Learning Curve
75% hand assembly/25% machining	80%
50% hand assembly/50% machining	85%
25% hand assembly/75% machining	90%

Table 9.2 NASA Learning Curves Based on Industry

Industry	Learning Curve
Aerospace	85%
Shipbuilding	80–85%
Complex machine tools	75–85%
Repetitive electronics manufacturing	90–95%
Repetitive machining	90–95%
Repetitive electrical operations	75–85%
Repetitive welding	90%
Raw materials operations	93–96%

Finding Your Learning Curve

Let's consider an example in which we wish to calculate a learning curve for a new assembly. Suppose the organization manufactured earlier model assemblies in the past. Historical data show that the company produced 1,000 assemblies at total cost of 695 hours, and that the first 500 assemblies cost 428 hours. With this information, the company can calculate its learning curve and the theoretical first unit cost. Here's how it's done:

- The average cost for the first 500 assemblies is 428 hours divided by the 500 assemblies, or 0.856 hours per assembly.
- The average cost for the first 1,000 assemblies is 695 hours divided by the 1,000 assemblies, or 0.695 hours per assembly.

Using this information in the learning curve equation yields:

$$0.856 = (C_1)(500)^{-P}$$

and

$$0.695 = (C_1)(1000)^{-P}$$

With apologies in advance for a foray into a bit of math, these are simultaneous equations that can be used to find p by dividing one into the other, as shown next:

$$(0.856)/(0.695) = ((C_1)(500)^{-P})/((C_1)(1000)^{-P})$$
$$1.232 = (0.5)^{-P}$$

The value for p can then be found by taking the logarithm of both sides, as shown here:

$$\log(1.232) = -p(\log(.5))$$
$$0.09049 = -p(-0.30102)$$
$$p = .09049/0.30102 = 0.3006$$

Based on this, the learning curve percentage is 2 raised to the negative p, or $2^{(-0.3006)}$, which equals 81 percent. We use 2 because the learning curve rate applies to each doubling of the cumulative production quantity.

As a further check, we can calculate the theoretical first unit cost two different ways. We can find this cost in the learning curve equation using the calculated value of p and the average cost of the first 500 units. Then we can do the same thing using our calculated value of p and the average cost for the first 1,000 units. If the answer for the theoretical first unit cost is approximately the same using both approaches, we know that the calculated values are good. The learning curve equation, rewritten to find the first unit cost, is:

$$C_1 = CC/(N^{-(p)})$$

Using both the first 500 and first 1,000 quantities and the average cumulative costs for these two quantities yields:

$$C_1 = 0.856/500^{-.3006} = 5.54 \text{ hours}$$
$$C_1 = 0.695/1000^{-.3006} = 5.56 \text{ hours}$$

The two numbers essentially agree (the small differences are attributable to variability in the cost information for the two samples). Using 5.55 hours as the theoretical first unit cost is a good approximation.

Armed with this information, the organization knows what its learning curve is, and it can prepare a table or a graph showing what the cumulative average cost should be for any amount of assemblies to be manufactured. This is easy to do using a spreadsheet, the learning curve equation, the theoretical first unit cost, the p exponent, and the learning curve equation, as shown in Table 9.3.[1]

---

1. The charts and tables included in this chapter can be downloaded at www.ManufacturingTraining.com. The Excel template also includes features for calculating learning curve percentages, theoretical first unit costs, etc.

Table 9.3 Assembly Learning Curve Predictions as a Function
of Quantity Produced.

Quantity	First Unit Cost	p	Average Cumulative Cost (hours per unit)
500	5.55	0.3006	0.857
1000	5.55	0.3006	0.696
1500	5.55	0.3006	0.616
2000	5.55	0.3006	0.565
2500	5.55	0.3006	0.528
3000	5.55	0.3006	0.500
3500	5.55	0.3006	0.477
4000	5.55	0.3006	0.459
4500	5.55	0.3006	0.443
5000	5.55	0.3006	0.429
5500	5.55	0.3006	0.417
6000	5.55	0.3006	0.406
6500	5.55	0.3006	0.396
7000	5.55	0.3006	0.388
7500	5.55	0.3006	0.380
8000	5.55	0.3006	0.372
8500	5.55	0.3006	0.366
9000	5.55	0.3006	0.359
9500	5.55	0.3006	0.354
10000	5.55	0.3006	0.348

The data from Table 9.3 are shown graphically as a learning curve in Figure 9.2. This x-y plot shows the characteristic shape of the learning curve. Note that as quantities increase, the rate of the cost reduction decreases, but the learning still continues.

This is useful information. With it, the organization can predict future cost reduction as a result of expected learning. If the organization is not realizing these anticipated cost reductions as the production quantity increases, something is wrong. The organization is not achieving the learning-curve based cost reductions that massive experience in other organizations indicates it should.

Note that in Figure 9.2, the projected (and expected) cost reductions are not linear. The rate of the decrease is significantly sharper earlier in the production run, and it becomes much less sharper as the production quantity increases. This is logical, as most of the learning should occur at the beginning. Gains continue as the production quantity grows, but at a smaller rate.

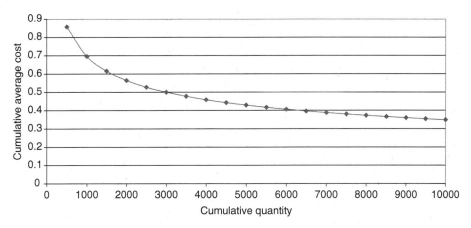

Figure 9.2 Assembly learning curve.

Factors Influencing the Learning Curve

As indicated at the beginning of this chapter, learning curves will vary based on several factors:

- The nature of the work being performed has a big impact. Work that is more labor intensive will see costs decrease more sharply. Less learning can be expected to occur in more automated operations; more can be expected to occur in labor-intensive operations. The NASA data shown earlier support this.

- Make-versus-buy content influences learning within the organization. More work done in-house will result in more learning occurring in-house. This does not mean that similar learning will not be occurring at suppliers. Prudent organizations should use learning curve theory as a negotiating tool for securing lower supplier prices.

- Training can have a huge impact on learning curve performance. As more efficient methods are developed and communicated to the workforce, the learning curve will become steeper (costs will decrease more rapidly).

- Process-improvement activities will improve learning curve performance. As tooling, setup, yields, and methods improve, costs will decrease more sharply.

- Sustained production and larger production rates will improve learning curve performance. This is logical, because it is more likely that improvements will progress during a steady run. Starts and stops in production can interfere with steady progress along the learning curve.

Perhaps the biggest factor influencing learning curve performance is management, and how well management communicates its expectations. If management is satisfied with existing costs, costs are less likely to decrease. If management is aware of the learning curve concept and track costs to determine if they are meeting learning curve projections, the costs are more likely to do so.

Applying Learning Curves

Our advice for using learning curve theory is simple:

- Develop a learning curve for every product currently in production and for those anticipated to enter production. It doesn't matter if it is just entering production or if it has been in production for years. Learning curve theory applies to both situations. It's not that hard and it takes little effort to do this.
- Compare actual costs to learning curve projections. Doing this communicates to the organization that cost reductions are expected.
- Where variances exist, find out why and act to eliminate the factors that are preventing conformance to the learning curve prediction.

Some may argue that if learning always occurs, an organization could relax, allow learning to occur on its own, and enjoy the subsequent cost reductions as they develop. Such learning will occur, but not as quickly as it would if the organization is actively seeking cost reductions. It's best to make the expectations known and strive to meet them, rather than simply letting nature take its course.

Over time, the organization will become more adept at accurizing its learning curve projections, using the learning curve as a tool for guiding expected cost reductions, and gaining confidence in the technique. As new business opportunities emerge, this expertise and confidence can be put to work when developing estimates for future production quantities, thereby gaining a significant cost advantage.

Who Should Do This Work

Developing learning curves and using them to predict future costs are tasks normally assigned to the manufacturing engineering group. If your organization does not have a manufacturing engineering group, or if it does and the manufacturing engineering group is not using learning curve theory, the application of this concept will require an in-house champion. A good approach is for the cost-reduction team to develop the initial learning curve and then train others in the manufacturing, engineering, and finance areas to do this work.

Risks

The relative complexity of learning curve mathematics may scare people away. Admittedly, the math is more complex than that required for the other concepts covered in this book, but it is not that complicated. It is further simplified through the availability of the downloadable spreadsheet mentioned in the introduction to this book (you can download the spreadsheet at www.ManufacturingTraining.com).

If the organization uses learning curve theory to predict future production costs and it wins business based on these predictions, the learning curve has to be met or profits will be less than anticipated. Predicted learning curves can be met if the organization actively evaluates its performance against the predicted learning curve and take steps to address variances if they occur.

The organization might predict learning curves that are too steep or too shallow. This risk can be mitigated by assessing past performance and comparing learning curve predictions derived from past performance against the NASA recommendations shown earlier.

References

C.J. Teplitz, *The Learning Curve Deskbook*, Westport, Quorum Books, 1991.
W.J. Stevenson, *Production/Operations Management*, Scarborough, Irwin, 1996.

Part II

Material

Material is another area with huge cost-reduction opportunities. Supplier purchases frequently make up 60 percent or more of a manufacturing organization's total costs, and an area this big can yield big savings. This section addresses several material cost-reduction topics, including:

- Make-versus-buy determinations.
- Inventory minimization.
- Material utilization.
- Minimizing supplier costs.
- Supplier negotiations.
- Supplier competition.

10

Make-versus-Buy Determinations

The Bottom Line

At the outset of a new program, organizations need to determine if they will make or buy required goods and services. Organizations should also periodically evaluate make-versus-buy decisions on existing programs. Bringing purchased goods or services in-house can sometimes reduce costs significantly; similarly, buying goods or services normally done in-house can sometimes reduce costs significantly. Net present value analysis is the preferred approach for quantitatively assessing costs and selecting the lowest cost approach. In addition to cost, risk is a significant factor in a make-versus-buy determination. The organization should identify all risks associated with a make-versus-buy decision and take steps to mitigate these risks.

Key Questions

What is our current make-versus-buy mix?

How do we make decisions to do work in-house versus buying goods or services from outside suppliers?

Are there any areas in which changing the existing make-versus-buy decision will result in significant savings, and do the savings outweigh the risks?

The Make-versus-Buy Road Map

When does it make sense to buy a product or service from a supplier, and when does it make sense to bring the work in-house?

These decisions are generically referred to as *make-versus-buy decisions.* Make-versus-buy decisions should be driven by the two factors of cost and risk. This chapter explores how to make a decision to do work in-house versus buying it from someone else.

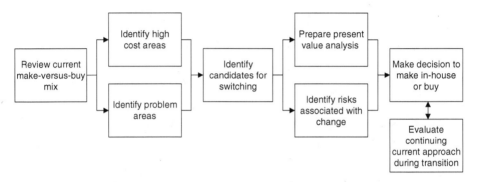

Figure 10.1 Recommended make-versus-buy approach.

Sometimes make-versus-buy decisions are so obvious that they require no analysis. No car manufacturer makes its own tires. It's cheaper and easier to buy them from the tire manufacturers. Other components are similarly obvious buy items, and it makes no sense to consider developing an in-house capability. This includes things like ball bearings, electric motors, electronic components, etc. There are organizations that specialize in doing this, and they can do it for less than it would cost to develop an in-house capability.

Apart from these obvious examples, though, there are many areas where the decision is less clear, and more detailed analysis is required. In many organizations, the make-versus-buy analysis is often not as rigorous as it should be, or the business situation changes after the analysis has been completed. Because of this, revisiting the existing make-versus-buy approach can offer significant cost-reduction opportunities.

Let's consider a few typical situations a manufacturer might face:

- Suppose your organization has a machining center that includes CNC equipment. If a milling machine breaks down, would it be better to contract with an outside repair service, or would you be better served if you had an in-house repair group?

- Suppose you manufacture fuel injectors for small engines, and the fuel injector bodies are zinc castings. Should your organization focus on design and assembly and buy the zinc castings from a casting house, or should you vertically integrate and cast the zinc parts in-house?

These are typically decisions that are made at fairly high levels in the organization. The decision typically involves cost and risk comparisons (as stated earlier), and can sometimes involve strategic considerations as well.[1] Although the final make-versus-buy determination may sometimes occur at higher levels, that doesn't preclude assessing the alternatives and providing appropriate recommendations.

A make-versus-buy cost assessment should be made in the same manner as a capital investment decision. Net present value analysis is the preferred approach for doing this. Net present value analysis is a technique that considers the present value of all income streams and all costs. It recognizes the time value of money and incorporates it into the analysis. Excel's present value function makes this easy to do, as we'll illustrate shortly.

A Typical Make-versus-Buy Decision

Consider the situation described earlier, in which a fuel injector manufacturer has to decide if it should purchase zinc castings for the injector bodies or make the parts in-house. The manufacturer wants to determine which approach presents the lowest cost. As part of this analysis, the manufacturer collects the following information:

- The manufacturer plans to continue making fuel injectors for the next ten years.
- The manufacturer sells 100,000 fuel injectors each month.
- Each fuel injector requires one zinc body casting.
- The casting supplier delivers their products to the manufacturer in monthly deliveries of 106,000 parts. The casting supplier delivers more than 100,000

---

1. For example, on large projects with global customers, manufacturers may opt to buy components from other countries as part of an agreement in which that country buys the completed product. In another example, defense contractors often seek to secure subcontractors in as many congressional districts as possible to secure and sustain congressional support for their products. On the other hand, some manufacturers keep most or all of their work in-house for brand differentiation and marketing reasons, even though subcontracting portions of the work would offer a cost advantage.

parts per month because about 6 percent are dimensionally nonconforming and cannot be used.

- The defective parts are returned to the casting supplier for a material credit of $0.14 per part.
- The manufacturer pays the casting supplier $1.38 for each conforming casting.
- The manufacturer estimates that if it makes the castings in-house, the cost of the zinc will be $0.12 per part.
- The manufacturer estimates that its purchasing agent spends 20 percent of his time overseeing the casting supplier. If the manufacturer opts to make the castings, less purchasing oversight will be required. The company estimates that if it makes the castings, approximately 5 percent of the purchasing agent's time will be required to buy zinc material.
- The purchasing agent's monthly cost, including salary and all benefits, is $12,240 per month.
- If the manufacturer opts to make the casting in-house, the cost of the casting equipment is $475,000.
- The cost of installing the casting equipment is $217,000, including all facilities' modifications.
- The cost of recruiting and training four operators to make the castings in-house is estimated to be $65,000.
- The cost of maintaining the casting equipment is estimated to be $25,000 per year.
- The cost of power for the casting equipment is estimated to be $152,000 per year.
- The monthly cost of each casting equipment operator, including all benefits, is estimated to be $4,440. The company would need four operators if it takes the casting operation in-house.
- The monthly cost of inspecting the castings is expected to be $2,589.
- Other administrative costs associated with the casting operation are estimated to be $2,100 per month (this cost is for inspection supervision, additional lighting, etc.).
- The useful life of the casting equipment is ten years. At the end of the equipment's ten-year life, its salvage value is $40,000.
- The cost of money (i.e., the interest rate, as provided the manufacturer's finance department) is 6 percent.

The challenge here is to decide if the company should make the castings or continue to buy them. As explained earlier, the net present value approach is the preferred approach for making this determination. The net present value approach brings all costs back to their present values. The sum of the present value costs for making the parts in-house can then be compared to the sum of the present value costs for buying the parts, and the manufacturer can assess the make-versus-buy decision from a cost perspective. The appropriate choice is the one that presents the lowest cost to the company (assuming acceptable risk).

As mentioned earlier, Excel's present value function makes this easy to do. The function is:

$$= PV(rate, nper, pmt, fv, type)$$

where

rate is the cost of money. If an input is an annual amount, the annual interest rate is used. If the present value of an input is a monthly amount, the interest rate is divided by 12.

nper is the number of periods. If the input is an annual amount, the number of periods is equal to the number of years. If an input is a monthly amount, the number of periods is the number of years multiplied by 12.

pmt is the monthly or annual amount.

fv is the future value of any known future amount.

type indicates whether any periodic payments are made at the beginning or the end of the period. Zero (0) indicates the payment occurs at the end of the period, and one (1) indicates the payment occurs at the beginning of the period. If the entry is left blank, the program assumes the payment occurs at the end of the period (as is normally the case).

In the cell where a present value is desired, we simply type = PV(...), inserting the appropriate parameters in the present value argument when we type it in the cell. Figure 10.2 shows what this looks like for a present value calculation.

24							
25							
26							
27							
28						=PV(6%,10,25000,0,0)	
29							
30							
31							
32							
33							
34							

Figure 10.2 Using Excel's present value function. When the Enter key is pressed, the above cell will show the present value of a ten-year series of annual $25,000 payments if the cost of money is 6 percent.

This is easier to understand by examining the net present value analysis for the fuel injector casting make-versus-buy decision. Table 10.1 shows the various costs and incomes, when they occur, and the present values for each.[2] The following paragraphs explain the present value calculations.

- The present value of the buy alternative's casting equipment, installation, recruitment, and training costs is zero. The fuel injector manufacturer will not incur these costs if it continues to buy castings from a supplier. If the fuel injector manufacturer opts to make the items, however, the costs are as shown in the table under the "Buy" column. Note that these costs do not need to have the Excel PV function applied to them. They are costs incurred at the start of the project, and as such, they are already at their present value.

- The salvage value for the buy option is zero. Salvage value is the value of the casting equipment at the end of its useful life. There is no equipment for the fuel injector manufacturer to sell at the end of its useful life if the manufacturer buys the castings from a supplier. For the make option, however, the fuel injector manufacturer will have to buy casting equipment, and as stated earlier, its salvage value at the end of its ten-year life is $40,000. The salvage value is given by the Excel present value function as:

$$= PV(6\%, 10, 0, -40000, 0)$$

The interest rate and number of periods are still 6 percent and ten years, but there is no monthly or annual payment (hence the zero for the third parameter). There is a salvage value, and that's the $40,000 the fuel injector manufacturer will receive for the equipment in ten years. We show it in the fv parameter spot (future value), and we show it as a negative number to indicate that it is the opposite of a cash outflow (it is a cash inflow because it will be a payment made to the fuel injector manufacturer).

- The annual maintenance and annual power costs are similarly taken back to present value for the make option. They are left as zero for the buy option because the casting supplier will incur these costs, not the fuel injector manufacturer. The fuel injector manufacturer's annual maintenance costs (for the make option) are:

$$= PV(6\%, 10, 25000, 0, 0)$$

That's because the annual interest rate is 6 percent, the number of years is ten, the annual payment amount is $25,000, there is no future value, and the

2. The Excel template for this analysis can be downloaded at www.ManufacturingTraining.com.

Table 10.1 Make–Versus–Buy Net Present Value Analysis. In this Situation, the Make Approach's Present Value Cost is $5.1 Million, and the Buy Approach's Present Value Cost is $12.7 Million. The Make Approach Represents a Significant Cost Reduction If the Risk of Making the Castings In-house is Acceptable

Net Present Value Analysis							
Cost of Money					6.00%		
Time Horizon (years)					10		
	Cost	Quantity	Monthly Quantity	Cost per Component	Monthly Cost	Annual Cost	PV (Make)	PV (Buy)
Casting equipment	$475,000	1	–	–	–	–	–$475,000	$0
Installation costs	$217,000	1	–	–	–	–	–$217,000	$0
Operator training and recruitment	$65,000	1	–	–	–	–	–$65,000	$0
Salvage value	$40,000	1	–	–	–	–	$22,336	$0
Annual maintenance	–	–	–	–	–	$25,000	–$184,002	$0
Annual power	–	–	–	–	–	$152,000	–1,118,733	$0
Monthly operator costs	–	4	–	–	$4,440	–	–$1,599,705	$0
Inspection (buy)	–	–	106,000	$0.02	$2,120	–	–$190,956	$0
Inspection (make)	–	–	103,000	$0.02	$2,060	–	$0	–$185,551
Other admin	–	–	–	–	$2,100	–	–$189,154	$0
Castings	–	–	106,000	$1.38	$138,000	–	$0	–$12,430,137
Zinc	–	–	103,000	$0.12	$12,360	–	–$1,113,308	$0
Reject credit (make)	–	–	3,000	$0.12	$360	–	$32,426	$0
Reject credit (buy)	–	–	6,000	$0.14	$840	–	$0	$75,662

Table 10.1 (cont.) Make–Versus–Buy Net Present Value Analysis. In this Situation, the Make Approach's Present Value Cost is $5.1 Million, and the Buy Approach's Present Value Cost is $12.7 Million. The Make Approach Represents a Significant Cost Reduction If the Risk of Making the Castings In-house is Acceptable

Net Present Value Analysis							
Cost of Money			6.00%				
Time Horizon (years)			10				
	Cost	Quantity	Monthly Quantity	Cost per Component	Monthly Cost	Annual Cost	PV (Make)	PV (Buy)
	Percent	Monthly Cost	Adjusted Monthly Cost (5% or 20%* $12,240)					
Purchasing oversight (make)	5.00%	$12,240	$612	—	—	—	–$55,125	
Purchasing oversight (buy)	20.00%	$12,240	$2,448	—	—	—		–$220,499.81
Net Present Value							–$5,153,221	–$12,760,526

payments are made at the end of the period (so these last two parameters are zero). The annual power costs are similarly determined.

- The monthly operator costs for the buy option are zero, again because if the fuel injector manufacturer buys the castings, it will not need to pay operators to make them. If the fuel injector manufacturer opts to make the castings, the monthly operator costs for the make option are:

$$= PV(6\%/12, 12*10, 4*4440, 0, 0)$$

Notice that for this present value calculation we are doing something different for the interest rate and the number of periods. That's because the annual interest rate is 6 percent, but this is a monthly payment so we have to convert the interest rate to a monthly amount. Similarly, the number of periods is 12*10, because there are ten years with twelve months each. The monthly amount is $4,440, but there are four operators, so this becomes 4*4440. The future value and payment parameters are again set to zero. The present value of the monthly inspection costs and other administrative cost are similarly determined.

- The present value of the castings under the buy option is the series of monthly casting costs taken back to present value. If the fuel injector manufacturer buys the castings, the monthly cost will be 100,000 times the $1.38 for each casting, or $138,000. The present value is:

$$= PV(6\%/12, 10*12, 138000, 0, 0)$$

- The present value of the castings under the make option is simply the value of the zinc material ($0.12 per casting) taken back to present value. In this case, the manufacturer believes the reject rate will be cut in half (from 6,000 to 3,000 per month), as explained earlier. The monthly cost is therefore 103,000 times $0.12, or $12,360, and the present value becomes:

$$= PV(6\%/12, 10*12, 12360, 0, 0)$$

- Under the make approach, the rejected castings can simply be remelted and recast, and the manufacturer allows for a positive monthly input accordingly. In this case, the monthly input is 3,000 times $0.12, or $360, and the present value becomes:

$$= PV(6\%/12, 10*12, -360, 0, 0)$$

- The present value of the rejected castings from the supplier is similarly calculated.

- Under both scenarios, a purchasing agent will either have to oversee purchasing the castings (for the buy option) or the raw zinc material (for the make option). The manufacturer has assumed that buying completed castings will occupy 20 percent of the purchasing agent's time, and buying raw zinc will occupy 5 percent of the purchasing agent's time. The present values are correspondingly calculated.

Once all of these calculations are complete, all of the present values for the make inflows and outflows are summed. The present values for the buy inflows and outflows are similarly summed. The Table 10.1 net present analysis shows that strictly from a cost perspective, the make option is much more attractive. The present value of the make option's cost is a little more than $5 million; the present the value of the buy option's cost is almost $13 million. It will cost the manufacturer significantly less to make the castings in-house instead of buying them from a supplier. If the manufacturer believes the risk associated with bringing the casting operation in house is manageable, it makes sense to do the work in-house.

Who Should Do This Work

Several departments within the organization should have input into make-versus-buy determinations:

- The finance group, the purchasing group, and the manufacturing group should take a lead role in identifying all costs and all risks associated with both making and buying goods and services.
- The quality assurance group can identify suppliers with poor quality performance.
- The purchasing group can identify suppliers with poor delivery performance.
- The sales group may have a say on customer perceptions regarding make versus buy determinations for different parts of the product. For example, certain parts of a product may be viewed by the customer as central to your company's niche in the market. For this reason, customer perceptions and marketing reasons may override the cost assessment.
- Senior management will have an input regarding risk-related decisions. In some companies, for example, key items influencing product performance are always done in-house to assure control over the process. Senior management makes these kinds of decisions to limit risk, even though the items can be purchased at a lower cost.

Risks

A make-versus-buy decision should not be based solely on cost. As explained earlier, risk is also a significant factor. In many cases, it may appear that making a component or developing an in-house service is less expensive then procuring the item or service outside, but any savings must be considered from a risk perspective. In other cases, it may appear that buying an item is less expensive than making it in-house, but the risk of buying it outside may be unacceptable. There are risks associated with making an item, and there are risks associated with buying an item. These risks must be identified and evaluated, in addition to cost, when making a make-versus-buy decision.

The most basic risk of all in deciding to make something in-house is the risk that you will not succeed. Suppose you decide to make an item instead of buying it and you are unsuccessful. This "What happens if I fail?" risk must be considered. One way to mitigate this risk is to pursue buying the item in parallel with the in-house development effort if the costs of doing so are reasonable. If the in-house approach fails, the buy option is still in place as a backup.

A related risk is labor and skills availability. A plan to make an item in-house requires assessing the likelihood that the in-house staff has the capacity and capability to do the work. If the capability and capacity do not exist in-house, there are risks associated with attempting to develop them.

There are also risks associated with buying an item rather than making it in-house. The risks here are that the supplier will not be able to deliver the item on time in the quantities required, or that the purchased items do not meet requirements. Using stable suppliers with good delivery history and strong engineering capabilities helps to mitigate these risks. Buying standard components also helps to mitigate the risk, assuming the standard components meet performance requirements.

Suppliers with proven histories offering standard components or services present minimal performance risk. Such suppliers frequently have other customers, however, and if you are a small part of their business, you may not have much pricing or delivery prioritization leverage. Again, multiple suppliers can help to mitigate this risk.

The risk of the product not meeting requirements increases when buying custom components or assemblies. This risk can be mitigated by clear and complete specifications, and performance testing prior to placing production orders.

Another risk with buy items is that the supplier will raise prices unexpectedly, or discontinue offering the product or service. Using multiple sources can mitigate this risk.

References

D.T. Koenig, *Manufacturing Engineering Principles for Optimization*, New York, Hemisphere Publishing Corporation, 1987.

J.R. Canada, *Intermediate Economic Analysis for Management and Engineering*, New York, Prentice-Hall, Incorporated, 1971.

11

Inventory Minimization

The Bottom Line

Minimizing inventory will reduce costs significantly. Manufacturers should not carry extra inventory for use as a cushion except in limited cases to prevent assembly-line shutdowns. Processes should be improved to prevention inventory inflation (i.e., making more parts than necessary) due to less than 100 percent yields. Infrequently used inventory should be sold or scrapped. The organization should track inventory turns and take actions to drive this metric higher. When material is rejected, the organization should require evaluation and disposition in less than twenty-four hours. The quest for high efficiency and utilization should not be allowed to inflate the inventory. The organization should keep its inventory secure to prevent shrinkage and misplaced items.

Key Questions

Do we have a program in place to reduce inventory?

Do we have any areas where we order more than needed to address process yield shortfalls?

Do we know how much obsolete inventory we have?

What are our inventory turns?

Do we check inventory levels before ordering?

How accurate are our inventory records?

What determines production lot sizes?

How do we handle rejected items, and how are we managing the effects on inventory?

The Inventory Reduction Road Map

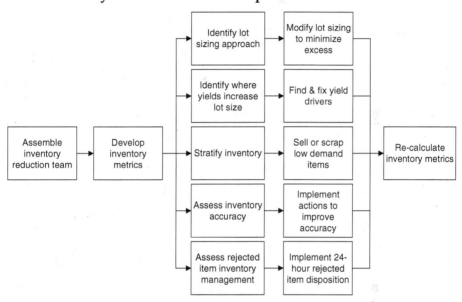

Figure 11.1 Recommended inventory reduction approach.

For the last few decades, the manufacturing mantra has been that inventory reduction is a good thing, and it's true. Costs come down with inventory reduction. Part of the reason for the cost reduction is obvious: Small inventories cost less than large inventories. Another reason inventory reduction reduces cost is that high inventory levels mask other problems, such as recurring nonconformances, scheduling errors, and excessive overtime. When the inventory level drops, inventory can't be used to hide these problems. The problems surface and have to be resolved, and doing so reduces cost.

Build-to-Stock versus Build-to-Order

The purpose of maintaining inventory is to have materials when needed. For manufacturers, this means having materials when needed to make parts. For sales organizations, this means having product available when orders are received. Some organizations stock items so they are available instantly. Even some manufacturers do this. These organizations build materials to stock, and the approach is called build-to-stock. Other manufacturers only produce product after receiving an order. This approach is called build-to-order. Build-to-stock organizations typically carry more inventory. In a build-to-order environment, theoretically inventory will only enter the organization when needed.

The reality is that both environments will carry inventories, but they will do so at different levels in the build cycle. A build-to-order organization, for example, may keep raw materials or semi-finished items in stock to minimize lead times. A build-to-stock organization keeps completed products in stock. In either environment, the objective should be to minimize inventory at all levels of the build process. Inherently, the build-to-order environment offers the potential for smaller inventories, but inventories will still be present, and the opportunity to reduce cost by reducing the inventory is therefore present as well.

Build-to-Stock and Build-to-Order Environments

In a build-to-stock environment, the organization is essentially betting that orders will come in and the inventory of finished goods will be sufficient to meet demand. The idea is that the company keeps an inventory of finished goods on hand so that when the customers want them, they are immediately available. If the finished goods inventory is larger than needed to meet demand, the inventory cost will be higher than it needs to be. If the inventory is lower than needed to meet demand, the inventory costs will be lower, but stockouts will occur and sales will be lost. There's a cost associated with that as well.

Addressing this issue in a build-to-stock environment requires examining the sales history, adjusting it for any outside factors (economic downturns, anticipated demand increases, etc.), and stocking work-in-process and completed inventory accordingly. This can be done either subjectively or quantitatively. If it's done subjectively, someone (usually someone at a fairly senior level) takes a guess at what the sales level will be. If it's done quantitatively, the approach involves defining sales history and adjusting the finished goods inventory based on seasonality, average demand, demand variation, and trade-offs between the likelihood of a stockout and the cost of carrying additional inventory. This is usually done by the sales or marketing department, with the final decision on how much material or product to keep in stock made by senior management. The problem with doing this either subjectively or quantitatively is that both approaches are dependent on forecasting, and business graveyards are littered with businesses done in by inaccurate forecasts. Even if a spreadsheet is used to apply advanced statistical concepts to the forecast based on demand variability and other factors, in the end the analysis is still based on somebody's guess as to what the sales level will be.

In a build-to-order environment, inventory levels at all process stages (raw materials, work in process, and finished goods) should be determined by what the next process stage requires. The finished goods inventory will be close to zero, because it is determined by exactly what customers have ordered, and the product should ship as soon as it is complete. Work-in-process inventory at each work center should be driven by exactly what the next work center needs. Raw material inventory should be driven by what the work centers that need the raw material require.

This concept of having only that inventory needed to meet the next work center's demand is called a *pull-based inventory approach*. This approach is most frequently encountered in the build-to-order environment, although it can also be used in the build-to-stock environment. In either situation, the next work center's demand "pulls" the inventory to it. This is a good approach, and it inherently minimizes inventory. Only those items required by the next operation are being pulled through the factory; there is no excess.

The opposite approach (and one that is most frequently found in a build-to-stock environment) is a *push-based approach*. It is driven by the sales forecast, anticipated process yields, and other factors that result in ordering raw materials and other supplier items, and then "pushing" them upstream through the factory.

From a cost-reduction perspective, the issue is what determines the inventory levels. If the organization is using a pull-based inventory approach, the organization is already focused on inventory minimization. If the organization is not using a pull-based inventory level approach, it may not be as focused. In either situation, there will be opportunities for cost reduction based on reducing the inventory.

Process Yield

Sometimes manufacturers make more parts then they need because they know some will be rejected. We saw an example of this in the last chapter, in which the fuel injector manufacturer ordered more castings than necessary because of an anticipated yield of less than 100 percent. This sort of thing occurs regularly, and the principal reason for it is that many people find it easier to order more inventory instead of finding and fixing the causes of the less-than-100-percent yield. They are using inventory to mask the problem.

The cost-reduction inventory opportunity here is obvious. Finding and eliminating the reasons for overbuilding requires determining the rejection's root causes. When this is accomplished, the extra inventory is unnecessary. This opportunity exists in any area where excess inventory is ordered to address yield problems. Chapter 19 addresses reducing cost by improving process yields.

Inventory Stratification

For a variety of reasons, the organization may have inventory that is no longer required or is required infrequently. Inventory stratification addresses this issue by assigning levels to inventory items based on demand. There are typically four levels ranging from A (frequently demanded) to D (infrequently or never demanded). Stratifying the inventory will probably reveal inventory-reduction opportunities.

Inventory stratification is easy. The organization's management information systems group can write a short program to assign a stratification level to each inventory item based on its usage history (the usage history is usually available, although many organizations do not use it). Management can define the criteria for an A level, a B level, etc. The good news is that it usually takes longer to decide what constitutes an A, B, C, or D level than it does to run the program and deliver the results. Once this information is available, the organization can then decide what to do with the items that are infrequently or never used.

It is counterintuitive, but it is usually better from a cost perspective to scrap unused or infrequently used items. Even if the unused or infrequently used items have been completely written off, there's still a cost associated with warehousing this inventory, recording it during periodic inventories, and other activities associated with keeping the items in stock. If unused items are allowed to remain in stock, there's also a subtle message from management that inventory minimization is not important.

Inventory Turns

The inventory turns metric is a method for assessing the "velocity" of the inventory compared to the total cost of making the product. The formula for it is:

$$\text{Inventory Turns} = \text{Cost of Goods Sold}/\text{Inventory Value}$$

The inventory turns calculation covers a specified time period (typically, the preceding twelve months are used for the cost of goods sold, and the inventory value is the value at the time the calculation is performed).

If the organization is not tracking the turns, that would be bad, but it means the opportunities for inventory reduction are great. If the organization is tracking turns but doing nothing to grow the turns, that would also be a bad sign, but again, the opportunities are great. In general, anything done to increase inventory turns is good because it inherently lowers the relative cost of the inventory.

In many organizations, the inventory turns are five or less, which is abysmal. Some companies have turns of less than one, which means that the inventory value exceeds the total cost of goods sold. That's worse than abysmal. Organizations using pull-based inventory systems can often attain turns of fifty or more.

Inventory Checks Prior to Ordering

This is so basic that it almost seems silly to list it, but the fact is many organizations order components or raw materials when the required items are already in stock. Sometimes orders are placed when there is enough inventory for the entire order (or more) already in stock. Sometimes there is a partial inventory on hand that would allow ordering smaller quantities. Checking the on-hand inventory before ordering more of the same reduces cost because it results in smaller order quantities and a smaller inventory.

Inventory Accuracy

Inventory accuracy refers to how well the inventory records reflect what is actually in inventory. Good accuracy is important, particularly when considered in light of the preceding topic. It's also important for another reason: If the inventory records show something is in inventory when it is not and the manufacturing organization is relying on that information, it won't have what it needs when it needs it. That can get very expensive, because it results in overtime, schedule disruptions, and other expensive problems.

If the organization does not know how accurate its inventory is, or if the organization knows the accuracy is low (98 percent is considered by many to be a minimally acceptable inventory accuracy), there are cost-reduction opportunities.

Checking inventory accuracy is relatively easy. It's usually done on a sampling basis. You can randomly select a few hundred inventory items in the organization's inventory records, and then check what's actually in the inventory. The accuracy is the number of correct records over the total number checked.

Inventory Inflation

In an earlier chapter, we pointed out that efficiency and utilization are useful for measuring the organization's performance and for targeting improvement opportunities. We also mentioned that focusing only on these two metrics could bloat the inventory. The reason for this is that in a quest to inflate the efficiency and utilization metrics, an organization can drive the inventory to higher levels than necessary. For example, in a drive to improve utilization the organization may make parts just to keep the machines running. If the parts aren't needed, it is just driving up the inventory and the overall cost. Similarly, an operator or a supervisor may keep a machine running longer than necessary to amortize the setup time over a larger numbers of parts. Efficiency increases, but the inventory swells.

The approach to identifying and eliminating this problem is to determine if it is occurring in the first place. It is easy to check production runs to determine if the operators are making more parts than the work orders require. It is also fairly straightforward to determine if the production schedulers are creating work orders that require more parts than needed to support the sales forecast or actual sales. Simply asking the schedulers how they determine work order lot sizes will reveal if this is occurring. If it is, it needs to stop. Making only the parts required will reduce the inventory, and in so doing, reduce costs.

Inventory Security

Inventory control refers to how well the inventory is secured, particularly in stockroom locations. For high-value, high-temptation products (for example, stereo systems in a motor home manufacturing facility), the concerns are obvious. In every organization, though, inventory security is important. People may not want to steal inventory items for personal gain, but they may need the items to mask production problems (rejections, lost inventory, etc.), or they may simply take inventory items for production without anyone knowing. When this sort of thing occurs, inventory accuracy suffers, and with it, costs increase due to material not being available when needed. The principal recommendation here is to keep inventory locations locked when an inventory control person is not in the location.

Inventory Storage

The importance of keeping work areas neat and orderly will be covered in detail in a subsequent chapter. To jump the gun a bit on that topic, we'll mention that keeping inventory areas organized, neat, and orderly is important for several reasons:

- Orderly inventory storage areas promote inventory accuracy.
- Orderly inventory storage areas prevent material from becoming lost.
- Orderly inventory storage areas prevent material damage.
- Orderly inventory storage areas make it easy to find required inventory items.

All of these reasons help to reduce cost. Figures 11.2 and 11.3 show examples of poorly-organized and well-organized inventory storage locations.

Figure 11.2 Improper inventory storage. This inventory storage area reveals several cost-reduction improvement opportunities, including inventory organization, security, and weather protection. Not surprisingly, this organization had recurring problems related to missing inventory.

Figure 11.3 A well-organized, clean inventory location. This stockroom, kept behind a locked fence, offers ready access to an organized inventory.

Rejected Inventory

Quality assurance and other inspections and tests occur throughout a manufacturing process. They start when supplier materials and components are received, continue during the in-house manufacturing process, and occur when the product is finished. The concern here is what happens to material when it is rejected. If not managed, rejected items can pile up in the work center, the inspection stations, or in special areas designated for rejected items storage. These rejected items increase cost for several reasons:

- The inspections and reinspections add cost.
- The rework (if the item can be reworked) adds cost.
- The cost of replacing the item (if it has to be scrapped) adds cost.
- The delay and disruption associated with taking the inventory out of circulation add cost. Many production scheduling systems do not recognize rejected versus accepted material status (they usually assume all material is accepted), so having rejected material on hand will present a false indication of what the organization has available.
- The administration of the rejected materials adds cost.

Reducing rejections offers significant cost-reduction opportunities. As mentioned earlier, Chapter 19 will address this topic. For now, we need to recognize that anything we do to reduce or eliminate rejections will significantly reduce cost, and anything we do to accelerate the disposition of rejected material will also significantly reduce cost. Our experience is that requiring rejected items to be processed within twenty-four hours of the rejection will reduce the cost of this inventory and minimize disruptions in the manufacturing process. Some companies require immediate disposition of rejected items, which also eliminates the need for rejected-item inventory storage areas.

Who Should Do This Work

Reducing inventory is not an easy task and accomplishing it requires inputs from many functions within the organization. The cost-reduction team should assist but not lead the inventory-reduction effort. We recommend establishing a separate team focused on inventory reduction, with representatives from the finance, manufacturing, material, management information systems, and quality assurance departments.

Risks

If the inventory is reduced too much, at some point the organization will need something it doesn't have. That can be expensive, especially if it stops production or costs a sale. The objective is to reduce inventory, realize the associated cost reductions, and do so in a manner that minimizes the likelihood of not having inventory when it is needed. The best way to do this is to eliminate the problems that created the high inventory level.

References

E. Goldratt, *Critical Chain*, Great Barrington, North River Press, 1997.
M. Muller, *Essentials of Inventory Management*, AMACOM, 2003.

12

Material Utilization

The Bottom Line

Material dropoff due to normal production operations can often be reduced by improved nesting approaches and use of near-net-shape premachining forms. The processes for applying liquids (paint, adhesives, and epoxies) can be improved to minimize material requirements. Dropoff materials, scrapped materials, and packaging materials can be recycled either in the plant or through recycling specialists. All of these approaches reduce cost by minimizing material requirements or by providing additional income streams. The organization should identify and implement suitable metrics for measuring material utilization.

Key Questions

What kinds of raw materials do we use in our production operations, and how do we maximize utilization of these materials?

Have we contacted any of our material suppliers and enlisted their support and expertise in minimizing material consumption?

What do we do with our dropoff, scrap, and supplier packaging materials?

Do we use any metrics for assessing our material utilization?

The Material Utilization Improvement Road Map

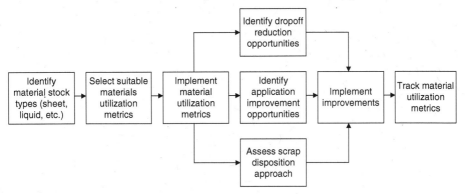

Figure 12.1 Recommended material utilization improvement approach.

Material utilization (not to be confused with labor utilization or machine utilization) refers to how much material is used versus how much is lost. In this context, we are addressing dropoff (material that is lost as a normal part of the manufacturing process), material consumption, and other factors that influence material loss. Material that is lost because it does not meet requirements (i.e., material rejected due to nonconformance) is addressed later in Chapter 19.

Examples of dropoff and excess material usage include:

- Chips cut during the machining process.
- Sheet metal or cloth that is left over after parts are cut.
- Remaining sections of angle iron, strap, I-beams, pipe, or other stock material after parts are cut.
- Paint overspray.
- Paint applied too thickly.
- Excess adhesive or epoxy.
- Injection-molded plastic flash.

The cost-reduction opportunity here lies in minimizing the dropoff and the amount of material consumed.

Machining

For machining operations, it obviously makes sense to start with material blocks that are as close to the final shape as possible (see Figure 12.2). For example, using pipe material as the starting point for a machined sleeve (if it is possible to do so) is much more economical than starting the machining operation on bar stock because the interior will not require as much machining. This reduces dropoff and machining time. If the finished sleeve dimensions are such that it cannot be made from pipe, perhaps the design can be changed so that it will allow machining the sleeve from pipe rather than bar stock.

If parts are to be milled from blocks, an approach that minimizes dropoff is to cut or purchase the blocks such that the outside dimensions of the block are at the outside dimensions of the finished part. Again, this reduces both material dropoff and machine time.

If complex shapes are to be machined, it makes sense to purchase a near-net-shape piece to start with (i.e., a shape close to the finished product) in a casting, forging, or molding such that material dropoff and machine time are reduced. This will require analysis to compare the costs of machining the part from stock versus buying near-net-shape castings, forgings, or moldings, but in many cases the use of near-net-shape blanks offer significant cost reductions in machine time and material.

Figure 12.2 Near-net-shape hydraulic actuator forging. Only minimal machining is required on this component, greatly reducing material usage and machining time.

Flat Stock

The layout of parts cut from sheet or roll stock (sheet metal, composite pre-preg, cloth, etc.) can be optimized to maximize the number of parts made from each sheet or roll. It doesn't make sense to use an entire sheet of material, for example, to cut out a single part that uses only a small portion of the sheet.

If your organization uses sheet stock or roll stock, we recommend looking into a layout nesting program. Several software packages optimize the nesting of parts to be cut from sheets or rolls.

In addition to optimizing parts nesting, the size of the sheets or the rolls can potentially be changed to reduce dropoff. Alternatively, the design might be modified to improve nesting such that dropoff is minimized. There are numerous cost-reduction opportunities here.

Standard Stock Forms

Many products use standard iron forms, such as warehouse racking and shelving, frames, jacks, etc. The production of these products typically involves cutting lengths from angle iron, pipe, I-beam, or strap. Manufacturers who use these materials order them in standard lengths. The challenge becomes optimizing the number of pieces cut from each standard length to minimize dropoff.

Occasionally, the mill providing the parts will agree to deliver nonstandard lengths so that there is no dropoff. The trade-off is the added cost for the non-standard length stock versus the cost of the dropoff.

As is the case for sheet stock described earlier, programs exist for optimizing cut lengths to minimize dropoff. In some cases, the material suppliers provide the programs at no cost.

The product's design should be assessed to determine if length changes would minimize the dropoff. Obviously, it does not make sense to make parts longer just to reduce dropoff (this would result in the same number of parts from the material), but if the parts can be made shorter, perhaps an additional piece could be cut from a standard length, resulting in no dropoff or reduced dropoff.

Paint

Any painting operation offers opportunities for minimizing *overspray* (which is lost material). Painting operations can also be optimized to minimize how much paint is applied to the product. Paint is expensive, and anything that can be done to minimize overspray or excess thickness will reduce cost.

Overspray and excessive coating thickness can be minimized by making adjustments to the paint mixture, selecting appropriate spray nozzles, adjusting pressure, using different processes (e.g., electrostatic paint application, powder coating, etc.), and training operators to use optimal spray patterns, techniques, and part-to-spray-gun distances. The technology for reducing paint usage is impressive, and it is usually available at no charge from the paint suppliers and the painting equipment suppliers. Paint producers and paint equipment suppliers provide support to help suppliers cut costs in this area, so if your organization has not used this resource it is missing a potentially significant cost-reduction opportunity.

In addition to training the painters and enlisting the support of paint producers and paint application equipment suppliers, it also makes sense to assign metrics and monitor the process with these metrics. Nondestructive inspection equipment for measuring paint thickness is readily available, and it allows rapid and easy paint thickness measurement (if the organization is not measuring finished paint thickness, it does not know how much paint it is applying). Another good approach is to track paint consumption and finished parts count. Dividing one into the other yields another metric for evaluating paint consumption. Paint thickness and paint consumption are both good metrics, and implementing one or both usually results in paint cost reduction.

Adhesives and Epoxies

Many manufacturing operations use adhesives or epoxies for joining parts. In this situation, it makes sense to work with the material supplier to determine optimal application thicknesses and optimal methods to prevent overapplication. Optimizing the thickness will improve product reliability (which will reduce cost), and optimizing application methods will eliminate using excess material (which will also reduce cost).

Many epoxies and adhesives have a shelf life (the *shelf life* defines how long the material lasts in storage). If inventories are not minimized, there's a good chance the organization will exceed adhesive and epoxy shelf lives, and the material will be useless. Taking steps to minimize inventory and keeping accurate inventory records (as outlined in the previous chapter) will minimize the likelihood of exceeding the shelf life. This will reduce cost by reducing the amount of material that has to be scrapped for exceeding its shelf life.

Epoxies have a pot life. *Pot life* refers to how long the epoxy can be used after being mixed (or how long it can be used after removal from the refrigerator if it is premixed). Production operations should be managed such that only the required amounts of adhesives and epoxies are ordered, only the required amount of epoxy is mixed or removed from refrigeration when it is to be used, and epoxies are used within their specified pot lives. Doing so will improve product quality by lessening the likelihood epoxies are used beyond their pot life, and by reducing the amount of epoxy that has to be discarded for exceeding its pot life.

Dropoff Recycling

Whenever dropoff occurs, it frequently can be recycled. In the case of machined parts or parts cut from metal sheet stock, scrap dealers pay for the chips and haul them away. This is a common practice and nearly any organization that cuts metal is already doing it.

Other dropoff and discarded materials may offer recycling opportunities. A few of the more novel ones are:

- Dropoff or scrap from molded or cast parts can frequently be added to the melt, lessening the amount of new material required.
- Aerospace manufacturers using composite materials can frequently sell their dropoff, excess material, and expired-shelf-life items to suppliers making components for less-regulated industries where the material can be used in nonsafety-critical applications. Manufacturers who make cosmetic carbon fiber accessories for motorcycles, for example, often buy discarded or outdated materials from aerospace suppliers.
- Cardboard recyclers will buy empty cardboard material from manufacturers and haul it away for free. Manufacturers who pay to have cardboard refuse hauled away as trash should investigate this option.
- Other packaging materials normally discarded after receipt can sometimes be recycled. These include plastics, wood, cardboard, steel, and other materials. One manufacturer even found that they could incorporate wood provided as part of their suppliers' packaging into their product.

Any of these approaches can reduce cost by either eliminating waste-hauling fees or by providing an income stream from the scrap material.

Material Utilization Metrics

Material is usually a significant cost for any manufacturing organization, so measuring material utilization with an appropriate metric will help to highlight, control, and minimize material costs. Here are a few suggested approaches:

- As previously mentioned, paint consumption and paint thickness are good measures for assessing paint material utilization. If the paint consumption metric is used, it should probably be normalized by dividing the amount consumed by the number of parts produced.

- Metals, lumber, pipe stock, and other similar materials can use metrics based on the amount of material consumed per month, although this should also be normalized by dividing the amount used by the number of parts produced.
- Scrap sales are a good metric. As in the previously mentioned case, the metric should be normalized by dividing the scrap sales amount by the number of parts produced.

Who Should Do This Work

The manufacturing, manufacturing engineering, and purchasing groups should take a lead role in material utilization improvements. The cost-reduction team can assist by recommending opportunities not currently being addressed by these groups.

Risks

The risks associated with improved material utilization are minimal. One possible risk is that in the quest to minimize dropoff the organization may make more parts than needed, but the inventory increase resulting from such an action is most likely negligible.

References

S. Kalpakjian, *Manufacturing Engineering and Technology*, Boston, Addison Wesley Publishing Company, 1992.

J.P. Tanner, *Manufacturing Engineering*, New York, Marcel Dekker, Inc., 1991

13

Minimizing Supplier Costs

The Bottom Line

Supplier costs are always a significant part of the total cost of goods sold, and for this reason, it makes sense to pursue supplier cost reductions. Specific actions to consider in seeking supplier cost reductions include relaxing requirements, incorporating savings-related design changes, developing "should cost" estimates, assessing suppliers' pricing history, asking for cost reductions, seeking quantity discounts, incentivizing suppliers to reduce cost, eliminating duplicate tests and inspections, using certified supplier programs, applying the learning curve, taking early payment discounts, and taking advantage of the allowed payment interval.

Key Questions

Have we asked our suppliers to tell us what we are doing that is driving their cost?

Have we prepared our own "should cost" estimates?

Are there any features on the supplier's parts that increase cost unnecessarily?

Do we ask our suppliers for cost reductions?

Does the supplier offer a cost reduction for larger quantity orders?

Do all of the requirements imposed on the supplier make sense?

Have we done anything to incentivize the supplier to reduce cost?

Is the supplier using the learning curve?

Are we duplicating any inspections or tests that the supplier performs?

Are we taking advantage of early-payment discounts and the allowed payment period?

The Supplier Cost-Reduction Road Map

Every business buys goods and services from other businesses. In most cases, supplier costs are more than 60 percent of a business's total cost of goods sold. For these reasons, keeping supplier costs down is one of the greatest opportunities for overall cost reduction. This chapter explores how to determine what a supplier's goods or services should cost and how to guide suppliers into meeting the "should cost" target.

Virtually every topic covered elsewhere in this book is applicable to suppliers. This chapter focuses on specific areas that can provide huge cost savings. The next two chapters after this one address supplier negotiations and supplier competition. Those two topics are specialized enough to warrant more focused discussions.

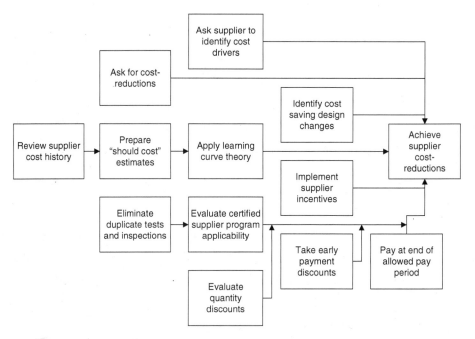

Figure 13.1 Recommended supplier cost minimization approach.

Requirements Relaxation or Elimination

Sometimes customers impose requirements on suppliers that make little sense and drive costs up significantly. Suppliers may recognize that the requirements are silly, but they might be hesitant to inform their customers. For example, when renewing a purchase order a company might impose first article inspection requirements on components that have been in production for years. If the new purchase order continues delivery of the parts, subjecting the parts to another first article inspection simply because they are being purchased on a new order adds unnecessary expense.

Labeling requirements may be imposed out of routinized engineering practice (the "we've always done it this way" syndrome). On many supplier components, labeling is unnecessary if the component comes in a labeled package. If a company buys small machined parts that are unique, packaged together in the same container, and cannot be mistaken for any other parts, it makes no sense to require a label on each part. Eliminating or simplifying labeling requirements is a good way to reduce costs on purchased items.

You should always ask your suppliers if any of your requirements add unnecessary expense. You should also ask yourself, your engineers, your quality assurance, and your purchasing people the same question. These questions frequently reveal what the cost drivers are, and in some cases, if they are unnecessary.

Design Changes

In examining supplier costs, you may discover features included in the supplier's products that raise cost unnecessarily. Here are some examples:

- If the supplier makes the part for others, it may contain features others need but you do not.
- The tolerances in purchased components may be unnecessarily stringent, either because they were imposed by your engineers or because they were imposed by the supplier. Tighter tolerances usually increase cost. Tolerance relaxations in supplier components may result in cost reductions.
- It might be possible for the supplier to use less expensive materials.

Discussing these design-related cost-reduction opportunities with the supplier can result in cost reductions. This book has a chapter on reducing cost with design changes. After reviewing that chapter, you can apply the same cost-reduction techniques to supplier products.

"Should Cost" Estimates

Your organization makes things, and it undoubtedly has the capability to define manufacturing processes and estimate costs. On high-value supplier components, it is a good idea to develop your own idea of what the product should cost. This information will be useful during the negotiation process (to be covered in the next chapter), but even before entering negotiations, it is a good idea to make this information available to your purchasing people. They need to have some idea of what the supplier component is worth prior to entering negotiations.

Purchasing History

Good buyers know the price history of what they are buying, and they will recognize unreasonable price jumps. Most manufacturing resources programs (MRP) and enterprise resources programs (ERP) maintain this history in their databases. Even in noncomputerized purchasing departments, good buyers will have buy cards showing the price history of each purchased item.

Sometimes the price history is revealing for other reasons. In one organization, prices from suppliers in completely different industries showed a consistent 6 percent annual increase. This organization had a rule that any price increase more than 6 percent had to be approved by the president, who was known to be a good negotiator. Price increases that did not exceed 6 percent could be approved by the buyers, and they typically accepted the increase without resistance. The company's suppliers all knew this, and the suppliers were ratcheting up their prices by 6 percent each year.

Cost-Reduction Requests

We'll talk more about this in the next chapter, but it is surprising how effective simply asking for a cost reduction can be. Some buyers simply accept the price the supplier quotes. Even when competing supplier offers are being considered, a good buyer will instinctively ask each supplier for a lower price. It doesn't hurt to ask, and good buyers always ask. In many cases, the suppliers will lower their prices. You just have to ask.

Quantity Discounts

Cost per item frequently decreases as quantity increases. It always makes sense to ask for a quantity discount. This does not imply that the organization should buy more than it needs (that would just bloat the inventory and raise costs elsewhere). You can usually get a quantity discount without having to receive the entire order in one delivery. You can ask suppliers for cost reductions on larger orders and have the supplier ship and invoice product as you need it. In other words, you buy a larger quantity and get the discount, but you take delivery and pay the same as if you were placing smaller purchase orders.

Supplier Cost-Reduction Incentives

Ordinarily, suppliers would not be very motivated to point out cost savings to their customers, because this would serve only to decrease the supplier's revenues or improve their customers' negotiating positions. Suppliers, especially in a competitive environment, are motivated to find ways to reduce cost when they are competing for business. After they win business, suppliers are still motivated to implement cost reductions, because doing so increases their profits.

As a customer, you want suppliers to find ways to reduce the cost for products they already have under contract with you, and you want to realize a cost reduction as a result. The challenge here is to get the supplier to want to reduce costs and pass some of the savings on to you. A great way to do this is to share the cost reductions with the supplier. Here's how this can work:

- The supplier proposes a design or process change that will result in a cost reduction. The assumption here is that the change requires your organization's approval; if it did not; the supplier could make the change, keep 100 percent of the savings, and continue to sell at the same price.
- The supplier explains the proposed design or process change and quantifies the estimated savings.
- If your organization approves the change, the supplier implements it.
- Your organization and the supplier split the savings. This is implemented by decreasing the supplier's price by half the savings. The supplier gets to keep the other half.

When this occurs, the supplier's overall revenues decrease, but the supplier's profitability jumps. Businesses usually prefer increased profits to increased revenues.

The U.S. Department of Defense has been using this approach with great success for decades (it's called their Value Engineering program). Some Department of Defense suppliers realize more than half of their total profits as a result of these cost-reducing design and process changes. It is a very successful approach to incentivizing suppliers to find cost reductions and share them with their customers.

Duplicate Tests and Inspections

Many times, tests and inspections performed at the supplier are also performed by the buyer, which makes no sense. The same tests and inspections don't need to be performed by both the buyer and the seller. Some organizations argue that they need to do this because the supplier is untrustworthy, but this is faulty logic (if this is the case, the organization needs to find another supplier).

It is better to eliminate the organization's in-house tests and inspections (and instead rely on the supplier's tests and inspections) if the product can be provided with documentation showing the test and inspection results. This may not result in a supplier cost reduction, but it should result in a cost reduction in your operation as a result of eliminating the redundant tests and inspections.

Certified Supplier Programs

This is closely related to the duplicate test and inspection issue previously mentioned. Certified supplier programs evaluate a supplier's quality assurance record. If the supplier has had no rejections in your receiving inspection, in-process inspection, and final inspection areas, some manufacturers opt to "certify" the supplier. This means that the supplier's deliveries go directly to your stockroom or the production line without the need for any in-house inspection. This can reduce your test and inspection costs significantly. Suppliers like it because they know it puts them in a preferred position for new orders.

Learning Curve Application

As explained in an earlier chapter, learning curve theory can help to drive costs down. If this supplier is using learning curve theory, it might be possible to work with the supplier to predict learning-curve-based cost decreases and realize additional savings.

Resisting Supplier Price Increases

Suppliers may unilaterally advise that they are raising their prices. Simply accepting the price increase would be wrong. In many cases, telling a supplier that you are not willing to pay the increase will make it go away. In other cases, it may lower the increase. Sometimes these unilateral increases can't be avoided if you are buying standard parts from a large supplier and you are a small part of their business, but in many cases, your organization can negate an increase by simply saying no to it. Buyer training is key to making this happen.

Taking Discounts

When suppliers invoice for their goods and services, the invoices typically contain phrases such as "2% Net 10." This means the organization receiving the invoice can take a 2 percent discount if the invoice is paid within ten days. Suppliers do this because they want their money quickly, and their customers pay within ten days because they want the discount.

A 2 percent discount is significant. It is a far better deal than hanging on to the money for another twenty days. Ordinarily, a customer would pay thirty days after being invoiced. Customers wait until the end of the thirty-day payment cycle to get the interest on the money. If they pay thirty days after receiving an invoice, they can get thirty days of interest on the money. If the annual interest rate is 5 percent, the amount earned on the invoiced amount would be 5 percent of 30/365, or 0.41 percent. If the organization pays in ten days, it gets the 2 percent discount plus the interest it earns on the invoice amount for ten days (or 5 percent of 10/365), which totals 2.14 percent. Getting 2.14 percent is a lot better than 0.41 percent, especially if one considers the amount of money paid to suppliers over a year. It's a good deal. If any of your suppliers offer a "2% Net 10" payment option, you should take advantage of it.

On the other hand, if your suppliers initiate a "2% Net 10" discount, check their prices. If they suddenly jumped 2 percent, they are just offering a discount on an inflated amount to get the money earlier, and you won't be saving anything. This should trigger a renewed negotiation or an effort to find another supplier.

Not Paying Immediately

As indicated in the previous section, the convention for paying invoices is thirty days after receipt, unless the supplier offers a "2% Net 10" discount option (in which case, you should pay after ten days). In either situation, it makes sense to hang on to the money and pay at the end of the allowed period to let the money earn interest (or to pay less interest if the funds are borrowed). If the interest rate is 5 percent, waiting thirty days to pay earns 0.41 percent, as explained earlier. That may not seem like much, but applied against millions of dollars, it's significant. If a manufacturer's annual payables are $10,000,000, for example, paying in thirty days versus paying immediately creates an annual savings of approximately $41,000.

Some manufacturers, realizing this fact, try to take advantage of a good thing and pay their suppliers in sixty or ninety days. We always advise against this for a number of reasons:

- Suppliers won't want to sell to organizations that pay late.
- Organizations that pay late get poor reputations, and they may lose customers based on a fear of imminent insolvency.

- Suppliers who continue to sell to slow-paying manufacturers will raise their prices or assess finance charges to cover the lost interest and administrative burden associated with the slow payments.
- Suppliers may impose a "cash on delivery" requirement when delivering their goods. This is a very difficult environment in which to work.
- Paying late is unethical.

Who Should Do This Work

The purchasing group has to take the lead in seeking the kinds of cost reductions reviewed here, with support from engineering, manufacturing, quality assurance, finance, and the cost-reduction team.

Risks

While suppliers expect their customers to pursue cost reductions, doing so carries a few risks. If a manufacturer is too aggressive in attempting to secure supplier cost concessions, the supplier may not wish to sell to that manufacturer. Good buyers know how hard to push and don't go beyond that point.

In seeking quantity discounts, manufacturers may order more inventory than is necessary. The approach for managing this risk is to commit to the largest order quantities that support production needs, and take delivery and pay only as the material is needed.

Eliminating in-house test and inspections that duplicate supplier tests and inspections is predicated on the supplier doing the testing and inspecting honestly and competently. This risk can be managed by supplier visits and assessing supplier history.

Sometimes critics of value-engineering programs (i.e., programs that incentivize suppliers to share cost savings) accuse suppliers of intentionally making the product more expensive in order to make money on cost reductions later. While this is theoretically possible, we have never seen it done. Nonetheless, if your organization feels a supplier is attempting to do this, the risk can be managed by a more stringent review of the supplier product prior to placing an order.

References

J.L. Bossert, (editor), *The Supplier Management Handbook*, Milwaukee, American Society for Quality, 2004.

R.A. Moore, *The Science of High Performance Supplier Management*, AMACOM, 2001.

14

Supplier Negotiation

The Bottom Line

Negotiating successfully with suppliers can provide huge cost savings. Successful negotiations require careful preparation and planning, a good understanding of the desired outcomes, and the ability to view suppliers as partners rather than opponents. Making the first offer, negotiating several issues simultaneously, and making well-planned, lower-priority concessions will help you achieve your objectives. Training can greatly improve an organization's negotiation skills.

Key Questions

How do we negotiate with suppliers?

Do we know our objectives and plan a strategy prior to entering the negotiation?

Who negotiates for us?

What has our history been in attaining what we want in a negotiation?

How skilled are our negotiators?

The Supplier Negotiation Road Map

This chapter offers approaches for planning and successfully negotiating reduced supplier costs. As mentioned earlier, purchased items usually account for 60 percent or more of any product's content, so reducing supplier cost is critical. Good negotiating skills are a key part of being able to do so.

The best time for using negotiation in reducing supplier cost is prior to placing a purchasing order, but it is not the only time. You can also go to suppliers on existing purchase orders and seek cost reductions.

Negotiation is a subject that many people find intimidating, because they imagine it involves confrontation (and most human beings don't enjoy confrontation). A successful negotiation does not involve confrontation. In fact, most negotiations that involve confrontation do not end well, because one or both partners in the negotiation are not satisfied with the outcome (even though they agreed to it).

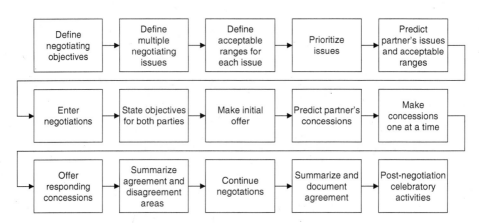

Figure 14.1 Recommended negotiation approach.

Being a good negotiator requires:

- Knowing how to define goals instead of positions.
- Negotiating multiple issues simultaneously.
- Knowing what you want the outcome to be.
- Knowing the range and limits of what constitutes an acceptable outcome.
- Knowing what your negotiating partner is likely to do.
- Timing and structuring concessions for maximum advantage.
- Tracking the give and take in a negotiation, and frequently reviewing it with your negotiating partner.

- Avoiding common negotiation traps.
- Remaining unemotional.
- Thinking as a problem solver, not a position taker.
- Knowing how to reach agreement.
- Knowing how to act after concluding negotiations.

Let's examine each of these topics, but before continuing, let's consider the "negotiating partner" phrase. It is important to think of whomever you are negotiating with as a partner and not an opponent. Most people tend to think in terms of defeating opponents, and that is not the objective of a successful negotiation. In a successful negotiation, both parties want to come away with a deal on the best possible terms. The objective in any negotiation is to get to this point while simultaneously getting what you want. When negotiating, if you think in terms of working with a partner rather than against an opponent, it's likely both parties will do better.

In line with this objective, successful negotiators define the desired negotiation outcome as a set of goals rather than a position on price. It often helps to review the goals with your partner at the start of each negotiation session. This emphasizes that the desired outcome is a situation in which both parties are satisfied. Doing this tends to lessen anxiety on both sides and foster a spirit of cooperation.

Planning

Good negotiations require careful planning. A good approach before entering any negotiation is to define what the desired outcome is in all areas and what the acceptable ranges are in each area. Good negotiators don't simply enter a negotiation, state a position on price, and see where it goes. They know what they want, and they have a plan for getting there.

Successful negotiators also prioritize the different areas in which they will be negotiating. They know what is important and what is not so important well before entering the actual negotiation. The not-so-important areas leave room for concessions that can be traded to gain ground in the important areas. This becomes part of the negotiating plan.

Good negotiators predict likely supplier responses, and have responses ready. This must also be part of the plan.

The First Offer

A question that always emerges when discussing negotiation is:

Who should make the first offer?

Many people are reluctant to make the first offer because they fear it will put them in a weak position. This is probably reinforced by what happens when purchasing a car, which is a negotiation scenario many people can relate to. Automobile sales folks are trained never to make the first offer (they always want you to make an offer first). These are people who negotiate for a living. The rest of us feel that if the car guys do this, making the first offer must be a mistake.

Actually, it is not. Negotiation studies have shown that negotiators who make the first offer usually come away with more favorable outcomes than their negotiating partners do. If we think about this in the context of the automobile purchasing scenario, the dealers always make the first offer, even though we believe they do not. The car's sticker price is the first offer. We have to respond to it if we wish to purchase the car.

One of the reasons most people fear making the first offer is that if their first offer is too generous, the negotiating partner will immediately accept it. When this happens, we feel like we've left money on the table. The key to avoiding this is to do your homework, understand what your negotiating partner (the supplier) is likely to accept, and include multiple issues with your first offer (with each having a value acceptable to you).

Don't be afraid to make the first offer on price. It should be below the lowest value you think the supplier will accept, but not so low as to be insulting. If you research the market and your negotiating partner well, your first offer will be interesting enough to make the supplier want to continue the discussion. That's when you can go on to the other multiple issues. On some of these less important issues, you know that you can make big concessions to get what you really want. Making the first offer increases the likelihood you will get it.

Multiple Issues

When most folks think of a negotiation, they focus on price as the only issue. That is not the best approach, even though obtaining the lowest possible price is an important outcome. The approach should be to simultaneously work multiple issues. The advantage of negotiating multiple issues at the same time is that it allows making concessions on less important issues. This often results in your negotiating partner making concessions that are critically important to you.

Here are just a few issues to consider incorporating as negotiation points when seeking a lower price with a supplier:

- Delivery schedule.
- Delivery quantity.

- Test and inspection approach.
- Follow-on orders.
- Cost-reduction profit-sharing arrangements.
- Supplier stocking requirements.
- Lead time.
- Delivery method.
- Warranty (items covered, length, exceptions, etc.).
- Labeling.
- Packaging.
- Sales exclusivity.
- Longer-term purchase orders.

Concessions

Making concessions is often perceived as a weakness, but successful negotiators know that selecting and timing concessions can be a source of great strength. Your most critical objectives may be to obtain a price 12 percent less than you are currently paying and a lead time under four weeks. Perhaps the other factors listed (delivery methods, packaging, labeling, etc.) are not so critical. A good strategy is to ask for things that might be difficult for the supplier to do in these noncritical areas, and then make sequential concessions on the noncritical items as negotiations progress while not conceding anything on price.

Good negotiators make concessions one at a time. They continue to ask for what they want in the areas important to them, and they wait for their negotiating partner to make concessions in those areas.

Tracking Progress

Good negotiators take copious notes during each negotiation session, and in particular, they note the concessions they have made. At the start of each negotiation session, reviewing the negotiation's progress (especially the concessions) emphasizes the negotiation's progress. This simultaneously highlights the concessions you have made and creates momentum. When done correctly, it can entice your negotiating partner to make concessions that bring you to your objectives.

Negotiation Traps

Several traps can derail a negotiation. Skilled negotiators know what these are and know how to avoid them. Here are a few, along with suggested avoidance approaches.

- One trap is becoming emotional. This never helps. If your negotiating partner becomes emotional, it is a good idea to take a break. Getting sucked into a confrontation is never good. Don't respond to temper tantrums. Remain unemotional and continue to focus on your goals.
- An offer to "split the difference" (e.g., agree on a price halfway between the two party's offers) is often a trap. The other party may have set an artificially high price. Splitting the difference may sound reasonable, but it often is not.
- Negotiating one issue at a time is a trap. Keeping several issues in negotiation allows maneuvering room.
- Ultimatums are a trap. The best way to respond to an ultimatum is to simply ignore it and continue the negotiation as if the ultimatum had never been made.
- Assuming (or being told) the other party has more power is a trap. You should treat the other party as a partner to be respected, but don't allow yourself to be cowed by perceived power or status differences.

Impasses

Good negotiators know how to solve problems in a collegial manner. Suppose you and your negotiating partner reach a point where it appears no further agreement is possible. A good way to approach this is to view it as just another problem. If you define the problem, identify the cause of the problem, identify solutions, and then pick the best solution, impasses frequently evaporate.

Summarizing the areas of disagreement and agreement with your negotiating partner is a good approach. When both parties see that they actually agree on many issues, it sustains the negotiation's momentum. When you ask your negotiating partner to help in resolving areas of disagreement (particularly when viewed in the context of the many areas in which agreement has been reached), the response is usually positive. Approaching the negotiation as a team endeavor makes this possible. Both parties want to reach agreement. If both work to find common ground, the results are frequently amazing. If you and your negotiating partner view an impasse as just another problem that you both can solve, solutions will emerge.

Negotiation Power

One of your strongest sources of power in any negotiation is the ability to walk away. This means having other options in the event your negotiating partner is not willing to agree to terms you find acceptable. If you need proof of this, consider what happens when you shop for a car. The dealers are usually much more willing to deal when you get up to leave. They know the probability of reaching a deal falls to zero once you leave the dealership.

We are not suggesting that you frequently threaten to leave a negotiation as part of your negotiating strategy. That's confrontational and it works against the objective of reaching agreement with your negotiating partner. We are suggesting that you have alternatives to reaching agreement with your negotiating partner. This places you in the very desirable position of being able to turn to your alternative solutions if the negotiation fails. You should not flaunt this ability, but you should delicately and diplomatically let your supplier know that alternatives exist. A good time to do this is at the start of the negotiation. You could mention, for example, that your objectives include reaching agreement with your negotiating partner on price, lead times, delivery schedules, etc., and that this is more preferable to you than having to seek alternative suppliers or bringing the work in-house. They'll get the message.

Conclude Negotiations Professionally

At the conclusion of the negotiation, good negotiators restate, document, and sign the agreed-upon terms. This cements the terms of the agreement and leaves no room for future misunderstanding.

Bragging about how you beat your negotiating partner down is, in a word, stupid. You can be sure word will get back to the supplier, and the relationship will be difficult (if not impossible) to repair.

Displaying goodwill after a negotiation is key to making the relationship stronger and sustaining the progress achieved in the negotiation. Good negotiators recognize they have done well for their organizations and they understand they have an ally and a business relationship. Parties, gifts, dinners, and other celebratory things are entirely appropriate, and they will make the relationship that much stronger.

Training

Negotiating is a skill that can be learned and developed, and a good negotiating class can pay for itself many times over. If you suspect your negotiators are not as skilled as they believe they are, our strong advice is to send them to negotiation school. Even skilled negotiators can benefit from training.

Who Should Do This Work

The obvious point of contact for negotiating with suppliers is the purchasing group. Others in the organization (quality assurance, manufacturing, engineering, etc.) can help the purchasing group by identifying areas as potential negotiation issues prior to the start of a negotiation. This will help support a multiple-issue negotiation.

Risks

As mentioned earlier, the purchasing organization should take the lead in negotiating with suppliers. The purchasing manager needs to make sure the buyer doing this is a skilled negotiator, and that he or she will be diligent in meeting the organization's objectives. Sometimes this puts buyers in a tough spot. Buyers are in boundary-spanner positions, and as such, they have to bridge the gap between the organization and the supplier. In many cases, buyers will have strong personal relationships with their suppliers. Sometimes the buyers' allegiances may be confused. Although a good negotiation results in both parties' satisfaction, your organization needs to make sure that your representative is looking out for your best interests.

Sometimes organizations bring more than one representative into a negotiation. They may do this to convey a sense of power or to provide support for the negotiator. Whatever the reason, only one person can speak for the organization. The other people from your organization should be there only for support when and if it is required. Prior to entering the negotiation, this needs to be clearly communicated to everyone. This extends beyond having just the organization's designated representative do the talking. The other team members need to avoid e-mails, text messages, phone calls, and other communication with the other party. They also need to avoid body language in the negotiating sessions. Some people have difficulty avoiding sighs, raising their eyebrows, smiling knowingly, rolling their eyes, or doing other things that communicate positions and sentiments nonverbally. Folks with this problem are usually more of a liability than an asset, and they should be kept out of the negotiation.

Sending someone into a negotiation who does not know how to negotiate is a serious mistake. Organizations should make sure that their negotiators are skilled in this area. Just because someone is a buyer does not automatically mean they are a skilled negotiator. If a buyer is a weak negotiator, the organization needs to send in someone else who is a strong negotiator. Too much is at stake.

Perhaps the greatest negotiating risk is going into the negotiation without being prepared. The approach outlined in this chapter will allow the organization to avoid this risk. Knowing the desired outcomes, predicting and planning responses to your negotiating partner's concessions, and going in armed with multiple negotiating points greatly increase the probability of success.

References

R. Fisher, and W. Ury, *Getting to Yes*, London, Penguin Books, 1981.
L. Thompson, *The Truth About Negotiations*, Boston, Pearson Education, 2008.
H. Cohen, *Negotiate This!*, New York, Warner Business Books, 2006.

15

Supplier Competition

The Bottom Line

Supplier competition is a powerful tool for reducing supplier cost. The organization should identify all areas where supplier competition is not used, evaluate where it makes sense to incorporate competition, and then do so. There are numerous risks associated with supplier competition. Careful analysis and planning are required to mitigate these risks.

Key Questions

Do we make suppliers compete for our business?

Where and why do we use sole-source suppliers?

Where does it make sense to incorporate supplier competition?

What are the risks that existing sole-source suppliers will exit the business if we compete the work?

What are the risks that existing sole-source suppliers will exit the business for reasons other than competition?

The Supplier Competition Road Map

Supplier costs can fall dramatically when suppliers compete for business. In many cases, merely mentioning a potential competitor will motivate a supplier to find ways to reduce cost. There are risks, however, associated with using competition as a tool for driving supplier costs down. This chapter describes how to effectively compete suppliers and presents a method for managing new supplier risk.

When seeking to reduce the cost of procured items through supplier competition, the first step is determining if your organization is already using competition. This may sound trivially obvious, but many organizations don't make suppliers compete for their business.

Sometimes there are good reasons for this, such as suppliers being sole sources for critical items. In this situation, it may not be possible to use competition to get the price down. If that is the case and the prices are satisfactorily low, it's time to move on to other opportunities.

If sole source suppliers have high prices, however, your organization has other options:

- You could reconsider the make-versus-buy decision.
- You could consider developing another supplier.

In many cases, simply informing sole-source suppliers that you are evaluating the development of alternative sources can produce results. We've seen several

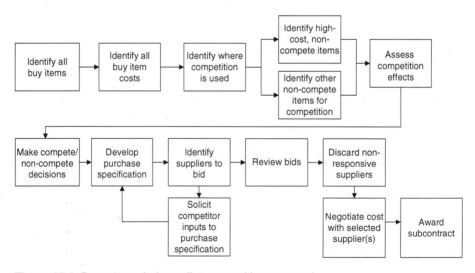

Figure 15.1 Recommended supplier competition approach.

cases where merely mentioning this induced sole-source suppliers to drop prices. Sometimes the price drop is significant.

The organization need not compete everything all the time. If you have a long-term relationship with preferred suppliers and their costs are low, competing the work may be counterproductive. If you never compete work, though, it's likely your supplier prices are higher than they need to be.

Implementing Supplier Competition

Here's a suggested approach for implementing supplier competition:

- Identify everything your organization buys.
- Identify the costs for everything your organization buys.
- Identify if competition is used when purchasing each of the items.
- Identify the highest cost items that are not competed. These are the items that probably offer the greatest potential for cost reduction, and these should be evaluated as potential items for competition first. This is a judgment call, and it may be that even the lower-cost items offer potential for competition-induced cost reductions.
- Evaluate the likelihood of competition reducing the price for each of the items identified as candidates for competition. If the item is a standard component used in many applications and you are a small part of the supplier's business (for example, if you are buying light bulbs from General Electric), it's not likely that competition will do much to drop the price. If the item is custom-made to your specifications, however, competition offers great potential for reducing the price.
- Identify the suppliers you want to bid the work. Ideally, you should know the suppliers well, even if your organization has not purchased from them in the past. Our recommendation is to solicit bids from suppliers with good quality and delivery records. There's no return in buying from a supplier who delivers low-quality equipment or delivers late just because they offer a lower cost.
- Develop a purchase specification that lists exactly what you require. The specification should be clear and offer no ambiguity. We recommend asking the potential suppliers to review the bid prior to finalizing it, soliciting their inputs, and where it makes sense to do so, incorporating the supplier inputs. The suppliers may recognize areas where unnecessary cost is built into the specification and advise you accordingly.
- Notify the selected suppliers, and evaluate the bids as they arrive. It's a good idea to reject any bids that are not responsive to the specification or that are late. If a supplier can't get a responsive proposal to you on time, their low prices will not overcome their inability to perform.

- After the bids arrive, ask the selected supplier for a lower price as part of the negotiation. Some people believe that in a competitive environment, the purchaser is obligated to accept the bids as submitted and make a purchase decision without further negotiation. We disagree with this stance. Everything can be negotiated.

Who Should Do This Work

The approach described here will require input from several groups:

- The cost-reduction team and the purchasing group should evaluate the list of purchased items to identify which of the sole source items make sense to bid. The chief executive may want to weigh in on this as well.
- The purchasing group will need to coordinate communications with the suppliers.
- The engineering, quality assurance, manufacturing, and manufacturing engineering groups will all have a hand in assessing the acceptability of potential bidders.
- The engineering group should prepare drawings and write the bid specification. The quality assurance group, the manufacturing engineering group, the production control group, and others will have to provide inputs for the specification.
- The engineering group, the manufacturing engineering group, the quality assurance group, and others will have to support the purchasing group in answering questions from the suppliers as the suppliers evaluate the bid specification.
- The engineering group, the manufacturing engineering group, the quality assurance group, and others will have to support the purchasing group in evaluating the bids.
- The engineering group, the manufacturing engineering group, the quality assurance group, and others will have to support the purchasing group after a supplier is selected to help the supplier satisfy the organization's requirements.

Risks

There are several risks associated with introducing competition where none previously existed. These are summarized next.

- One risk is that competition based solely on price (with no consideration given to any other factors) could result in the winning supplier defaulting after winning the work. If the supplier did not understand their costs, or if they intentionally decided to underbid the work and somehow make it up later (we've seen this happen), the supplier may be forced out of business or simply walk away from the work after realizing that it is financially unworkable. In that case, competition has put your organization in a terrible position. The approach to managing this risk is to know the suppliers you intend to buy from (which is wise advice in any circumstance). You need to assess the supplier's financial stability, and determine if their price is in line with your "should cost" estimate. You should also check references and other information sources to determine if the supplier has ever defaulted on an order. Another approach to managing this risk is to award the work to two qualified bidders, with the percentage split determined by their relative prices (a higher percentage would go to the supplier with the lower cost). This approach maintains multiple sources and mitigates the effects of a single supplier defaulting.

- If supplier competition results in the work going to an organization that has not built the item in the past, there is a risk in the supplier transitioning the job to production. The issues to explore here are the suppliers' engineering and manufacturing capabilities and their success in transitioning similar products to production.

- If the purchase specification you use when soliciting bids is not precise, there is a risk that the winning supplier will provide conforming product that doesn't meet your needs. If this occurs, you are stuck, and the supplier would be within their rights to demand additional funding if their product requires modification to meet your needs. To manage this risk, your specification has to exact, it has to be correct, and it has to be unambiguous.

- Another risk when introducing competition is that your existing sole-source supplier may decide to exit the business. This is a bit tougher to evaluate. You need to know the existing supplier and how they are likely to react. If the material the existing supplier provides to you is a major part of their business, they are probably not going to walk away from it immediately (although we've seen this happen, too). If your work is a very small part of their business, the existing supplier may lose interest and no-bid the work. In that case, you could be in a considerably worse position than you were before you introduced competition. Managing this risk requires predicting the existing supplier's likely reaction and objectively assessing the success of a new supplier. It's a subjective assessment, as is the case in many business decisions.

- There's a related risk to the one described in the previous bullet: A sole-source supplier could exit the business even without your having introduced competition. This can happen because of business failures, sale of the supplier to another organization, or the supplier may simply decide not to offer the product for profitability, product mix, or other reasons. This is a different risk, and it argues for introducing competition for reasons other than cost.

- Finally, there's a transition cost associated with moving to a new supplier, even if the supplier offers a lower cost for its product. The organization should identify and quantify likely transition costs prior to introducing supplier competition.

References

J.L. Bossert, (editor), *The Supplier Management Handbook*, Milwaukee, American Society for Quality, 2004.

R.A. Moore, *The Science of High Performance Supplier Management*, AMACOM, 2001.

Part III

Process Improvements

Process improvements are the third major area to be covered in this book. This broad category also offers cost-reduction opportunities in the areas of:

- Work-flow improvement.
- Setup time reduction.
- Material-handling improvement.
- Scrap and rework reduction.
- Work center cleanliness and organization.

16

Workflow Optimization

The Bottom Line

Processes should be designed such that workflows in a linear manner, there are no duplications of effort, distances between operations are kept as small as possible, and the need to leave the work center is minimized. A process flowchart and facilities layout diagram will frequently reveal improvement opportunities. Assembly lines should be designed with subassembly work areas located off the main assembly line. Work on complex assemblies should be designed so that operators approach the assembly with all required tools and materials and work around the assembly in a linear manner.

Key Questions

How do we design our processes?

Does our work flow in a straight line?

Have we flowcharted our processes to search for improvement opportunities?

Do operators have a defined sequence for what they do, or is it left up to them?

The Workflow Optimization Road Map

This chapter examines how to optimize workflows. Costs can be significantly reduced by eliminating unnecessary steps, by closing work-flow gaps, and by "straightening" the process (having the work progress in a linear manner rather than moving around the facility erratically). Some folks refer to this as "lean manufacturing" and talk about the concept as if it is something new, but it has been around for decades.

Workflow optimization means designing the process and the manufacturing facility such that:

- Workflows in straight lines.
- Distances from one work center to the next are kept short.
- Inventories and tools are located close to where they are needed.

The concept is that doing these things minimizes time away from production and transport time. For many reasons (most of the reasons are related to the existing facility), it will not always be possible to completely comply with the listed items, but the idea is that any move toward compliance is going to increase efficiency and minimize lost time.

Flowcharting and Process Analysis

Optimizing the work flow requires flowcharting the process to completely understand what it takes to make the product. This sounds obvious, and many organizations feel they already understand their processes, but experience shows that when a manufacturing team flowcharts the process, there are always surprises. The idea behind flowcharting is that it allows the organization to

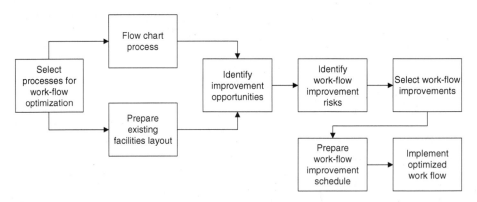

Figure 16.1 Recommended workflow optimization approach.

understand the process steps, where inventory feeds into the process, where tools are located, where machines are located, and so on.

When preparing flowcharts, the best approach is to observe the existing process, work with people who understand the process, and ask questions. It's best to ask a lot of questions and take notes, and then draw the charts by hand initially. As these initial hand-drawn flowcharts are created, asking those who do the work if the flowchart is accurate usually results in corrections, clarifications, and additions. We've always found that hand-drawn charts make people more willing to point out corrections. Showing someone a computer-generated drawing may make people reluctant to identify where it is incomplete or otherwise wrong.

After preparing the flowchart, the team should next prepare or obtain a drawing that shows the physical layout of the existing process, and compare it to the flowchart. This document similarly does not need to be computer-generated unless it already exists. Hand-drawn charts on large sheets are preferred, as they foster creativity and allow for easy changes during the analysis.

Questions the organization should focus on after preparing these charts are:

- Have we identified all of the steps?
- Are all of these steps necessary?
- Does each of the steps add value?
- Is the sequence logical?
- Are there points in the process where the work-in-process inventory backs up?
- Are there points in the process where work centers have to wait for product or inventory?
- What can we do to eliminate any bottlenecks?
- Are there points in the process where we have to undo work done in a previous step to allow access or assembly in a subsequent step?
- Where do we inspect the product, and are all of these inspections necessary?
- Where do the people in each work center have to go to get tools or inventory?
- What other things make it necessary for people to leave their work centers?
- What can we do to minimize the need for operators to leave their work centers?
- What can we do to make the work flow more logically?

- What can we do to minimize the product travel distance?
- Where are we doing the subassembly work?
- Where can we locate subassembly work offline so that it doesn't interfere with the assembly work?
- What can we do to get the inventory closer to where it is needed?
- If we were starting the process design with a clean sheet of paper, what would we do differently?

One high-technology example that illustrates this is the process for making ceramic jet engine fan blades for the new F-35 aircraft. Figure 16.2 shows the original flowchart for this process, with the inspection areas highlighted in yellow.

The process steps for this product include receiving raw materials, inspecting the raw materials, mixing plastic resins, baking the resin to flash off the volatiles, bagging and laying up the parts (putting them in their required shapes and enclosing these shapes in vacuum bags), baking the parts, inspecting the parts, machining the parts, inspecting the parts again, and then repeatedly infiltrating the parts with another material and baking the parts. The components go through the infiltration and baking steps several times.

As the team flowcharted the process, several opportunities emerged:

- The process contained numerous inspections. The team working this process concluded that several of the inspections were redundant. They also learned that no rejections were occurring at several of the in-process inspection points. The team opted to eliminate the redundant inspection points and all of the inspection points that had a history of zero rejections (except for the last one). Their logic was that the cost of the zero rejection inspection points was higher than the cost of rejecting parts at the end of the process. It was more economical to accept the risk of scrapping a few parts at the end of the line rather than constantly inspecting at each stage of the process, although the team also initiated an effort to find and eliminate the causes of the rejections.
- The initial material flow was horrendous, with numerous instances of work in process changing direction to get to the different work centers (see Figure 16.3). The team rearranged the flows so that the flow was more linear, with only limited instances of the work having to reverse direction (see Figure 16.4). The team found that difficult-to-move pieces of production equipment (these are sometimes called "monuments") could not be moved, so the flow could not be completely linear, but the rearranged flow was quite a bit better than the original flow.

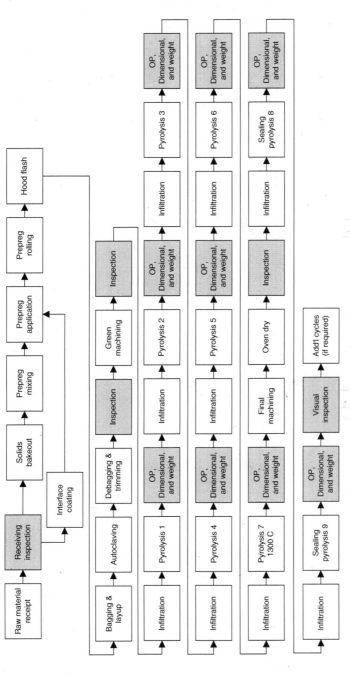

Figure 16.2 Jet engine ceramic blade manufacturing process flowchart. The flowchart shows all steps in the process, with the grey blocks indicating tests or inspections.

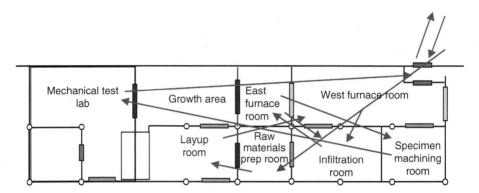

Figure 16.3 Jet engine ceramic blade production initial facilities layout. Note the convoluted path the materials must follow as parts are manufactured.

- The team found facility constraints. It was not possible to redesign the facility to keep everything moving in a straight line. In these cases, the work had to make turns, as Figure 16.4 shows. Even with these constraints, the team focused on keeping the work moving as linearly as possible, and keeping the distances between work centers as short as possible.

Assembly-Line Workflow Optimization

Sometimes products are built on an assembly line. This occurs for larger, more complex systems, such as aircraft, cars, and other similar large assemblies. In this case, the work flow is designed to create an assembly that contains numerous subassemblies. The challenge here is to keep the assembly line moving, because if it stops, the entire plant comes to a halt.

In these situations, it is still desirable to keep the work moving in a straight line, but a new issue emerges involving the subassemblies used in the larger assembly. It is a good idea to locate the subassembly work centers close to the assembly line, but not make these supporting work centers part of the assembly line. The reason for locating the subassembly work centers off the assembly line is that it prevents subassembly work stoppages from stopping the assembly line. If the subassembly work centers build very slightly ahead of the assembly line's need, the subassemblies are always available for the assembly line and there is time to correct the causes of any subassembly work stoppage.

As mentioned earlier, it makes sense to build subassemblies very slightly ahead of the assembly line's needs. This is counter to the concepts presented earlier regarding excess inventory, but in assembly-line situations, the trade-off of very slightly increased inventory is a logical trade-off when balanced against the cost of a plant shutdown.

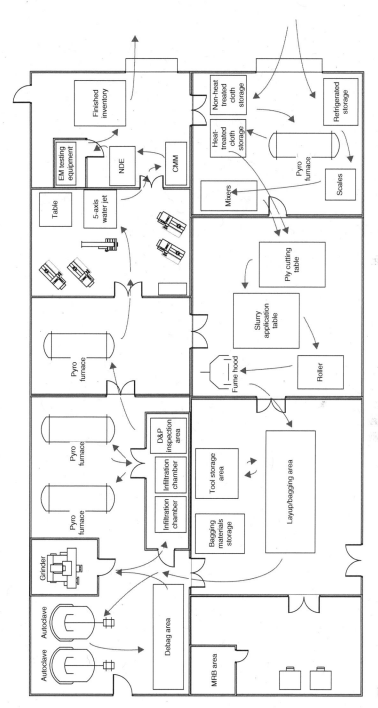

Figure 16.4 Jet engine ceramic blade production modified facilities layout. The process has been "straightened" to the extent possible, with the work flow moving in a relatively linear manner. Material enters in the lower right area of the figure, moves to the left, makes a u-turn, and then exits at the upper right area. Although this is not a perfectly straight line, it is far more linear than the initial work flow.

Within the subassembly areas, manufacturers should follow the same principles outlined in this chapter and elsewhere in this book. They should use linear work flows, locate the inventory and tools where they are needed, minimize distances, and do other things to reduce lost time.

Figure 16.5 shows a plant layout for manufacturing motor homes, which illustrates the assembly-line concepts outlined here.

Many assembly areas require multiple operations by several people in each assembly-line work center. In these situations, arranging the work so that it is performed in a progressive manner without the operators having to change directions, cross paths, or leave the product will reduce cost significantly.

The motor home manufacturer's initial approach was to simply define all of the steps each of several operators had to perform at each workstation on the assembly line. The result was inefficient; as operators crossed paths with each other, they had to reverse directions several times when moving around the motor home, and each had to get off the motor home several times to retrieve tools and inventory.

Based on a desire to eliminate these problems, the motor home manufacturer redesigned the process so that within each assembly line work center:

- Each operator performed tasks moving from one end of the motor home to the other in a continuous circle.

- Each operator had all required tools and materials as he or she entered the motor home.

- Operators followed each other as they completed their tasks without having to cross paths.

The savings were enormous. Under the first somewhat disorganized approach, each motor home required more than 1,700 labor hours. After redesigning and sequencing the operators' tasks as described here, the labor content per coach dropped to slightly less than 700 hours.

Who Should Do This Work

The cost-reduction team should work closely with the manufacturing, manufacturing engineering, purchasing, quality assurance, and material groups when redesigning processes and plant layouts as described in this chapter. In many cases, each of the groups mentioned here has other issues that occupy their day-to-day activities. A "big picture" perspective is required for this kind of work, and it usually needs fresh eyes to avoid "we've always done it this way" thinking. The cost-reduction team should take a lead role in facilitating the work-flow optimization effort.

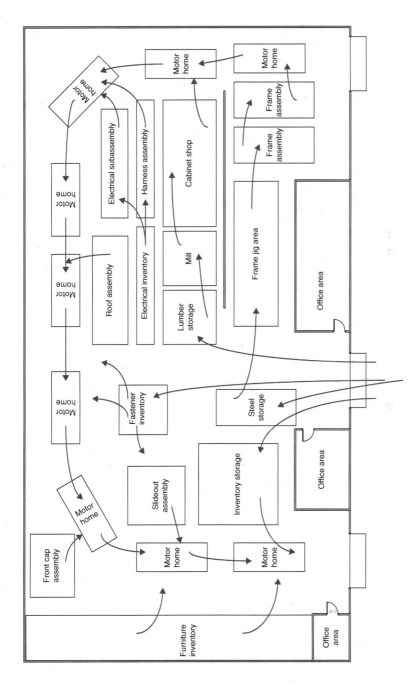

Figure 16.5 Motor home layout. Note how the motor home assembly line moves in a continuous line, with subassembly work centers located off the assembly line (but close to where they are needed to support the assembly line). Materials enter in the center front of the factory, motor home chassis (which are procured from the chassis manufacturer) enter at the right front of the factory, and completed motor homes leave at the left front of the factory.

Risks

The biggest risk related to work-flow optimization is that a redesigned process may not work as intended. In our experience, that risk is low. The organization can make it even lower by:

- Actively soliciting input from the operators, the manufacturing engineers, and manufacturing supervision when creating the flowcharts, facilties layout drawings, and the revised process.
- Asking how things can go wrong at each step of the revised process.
- Considering the physical dimensions and weight of the product to make sure that it can move and fit through the facility in accordance with the redesigned process.
- Making sure any required inputs (electricity, compressed air, environmental controls, etc.) are available where needed in the redesigned process.
- Making sure the redesigned process and its associated facilities' modifications comply with safety, zoning, hazardous material, emissions, and other regulatory or legislative requirements.

Another risk is disruption to the manufacturing process when the process changes are incorporated. There will be a disruption; the objective must be to minimize its effect. One option is to make the changes during the weekend or in the evening. Equipment and facilities modifications must be closely choreographed to minimize disruptions. The team needs to consider any equipment items that require recalibration, hookup, or other actions, and make sure these activities are scheduled and executed appropriately.

In assembly-line operations, having the subassembly work centers build slightly ahead of the assembly line's need mitigates the risk of an assembly-line stoppage, but doing so slightly increases the work-in-process inventory. The team can mitigate this risk by assuring inventory accumulations in the subassembly work areas are kept as small as possible consistent with preventing an assembly-line stoppage.

References

S.A. Ruffa, *Going Lean: How the Best Companies Apply Lean Manufacturing Principles to Shatter Uncertainty, Drive Innovation, and Maximize Profits*, AMACOM, 2008.
K.W. Daily, *The Lean Manufacturing Pocket Handbook*, Grand Blanc, DW Publishing, 2003.
J.P. Ignizio, *Optimizing Factory Performance*, New York, McGraw-Hill, 2009.

17

Setup Time Reduction

The Bottom Line

Setup refers to activities required to make the process ready to produce. Setup time reduction is highly desirable because it increases capacity and reduces cost. Setup times can be reduced through a variety of techniques. Wherever possible, setups should be performed offline to not interfere with run time (the time spent actually making product). The organization should rank its setups by time consumed and propensity to induce bottlenecks, and focus setup time reduction activities accordingly.

Key Questions

Do we have a program in place to reduce setup time?
Do we perform setups online or offline?
Which of our setups consume the most time?

The Setup Time Reduction Road Map

Setup time reduction results in huge cost savings for several reasons. It increases capacity, improves quality, improves work center and factory

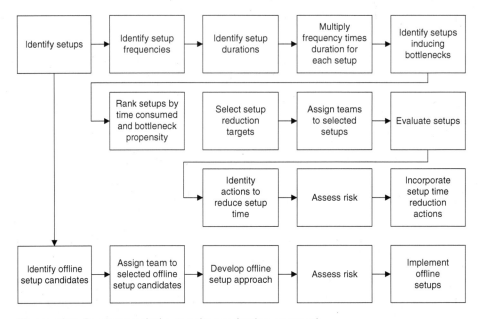

Figure 17.1 Recommended setup time reduction approach.

responsiveness, increases output, reduces inventory, decreases required floor space, and makes the plant more flexible.

The term *setup* refers to preparing, adjusting, and making ready the equipment, people, and facilities required for a manufacturing operation. Setup activities can include:

- Setting up machining equipment such as lathes, mills, and drills.
- Setting up molding or casting equipment.
- Setting up a paint line.
- Moving workbenches and people to new positions required by a particular product.
- Installing parts in fixtures for subsequent operations.

These are just a few examples. For our purposes, setup is anything we have to do with people, machines, tools, or other things so that we can make product.

The reason reducing setup time is so important is that it frees up people and equipment to make parts. In many cases, the setup time is longer than the run time required to actually make the parts. Whatever time can be taken out of the setup process is time that can be put to use making product.

Setup Time's Negative Impacts

Long setup times have several adverse affects:

- The first and perhaps most obvious is that setup time is lost time. When operators are engaged in setup activities, they are not making parts. Time spent in setup is time that is taken away from actually making parts. Lost time is a cost driver, and anything we can do to decrease it reduces cost.
- A second adverse effect (closely related to the previous one) is that setup time decreases the plant's capacity. Time spent in setup is time that people and machines lose for making parts, and this reduces capacity.
- The third adverse effect is a natural tendency to want to make more parts than the work order requires. Many organizations mistakenly believe they are better served by making more parts than they need to amortize long setup times over a larger number of parts. The intent is good, but the results are bad. Increasing the number of parts beyond what is actually needed is bad for two reasons – the setup time remains long and the organization's inventory swells.

Reducing Setup Time

In an ideal world, setup times would be zero. If that were the case, we could instantly switch from making one part to making a different part with no lost time (there would be no setup time required). In the real world, we understand that most of the time we won't be able to attain a zero setup time, but whatever we can do to bring the setup time closer to zero is inherently good. It reduces the tendency to make more parts than necessary, shortens lead times, and reduces lost time.

Here's the recommended approach for reducing setup times:

- Identify all processes in the facility requiring setup.
- For each setup, list how often it occurs (e.g., number of times per day or per week) and the approximate setup time (in minutes or hours).
- Multiply the frequency of occurrence by the time required for each setup.
- Sort these resulting times from highest to lowest. This will identify the largest sources of lost time due to setups.
- In addition to this prioritization based on setup times and frequencies, identify areas where setups induce bottlenecks in the production line.
- Based on this prioritization, select target setups for setup time reduction.
- For each selected setup, assign a team to focus on reducing that process's setup time.

Once setup reductions have been prioritized in the specified manner, several activities will help to identify how to reduce setup time. The people who actually do the setups should demonstrate it to the setup time reduction team. It helps if the team digitally records the setup for further study. When observing the setup, the team should identify where the setup requires:

- Waiting.
- Searching for parts, tools, materials, or drawings.
- Preparation activities.
- Making adjustments to tools or measuring devices.
- Loading parts into tools, or tools into machines.
- Unloading the previously listed items.
- Walking.
- Reading.
- Performing setup activities offline (i.e., in parallel with ongoing production, such that production time is not lost).
- Performing setup activities online (i.e., setup actions that prevent production from starting until the setup is complete).

Each of these areas will probably contain opportunities for reducing the setup time. After the team understands the existing setup, brainstorming is a good way to develop ideas. The team can document their ideas, discuss which ones offer the most promise, and then experiment to see if the ideas work.

There are a number of specific techniques for reducing setup times. The following discussion suggests several approaches for reducing setup times.

Fixturing

Fixtures are tools that hold parts in position so that they can be assembled, machined, painted, or subjected to other operations. Whenever fixtures are used, it is necessary to locate and then lock the part or assembly being fabricated into position. Anything that accelerates securing or removing the part in the fixture will shorten the setup time. Here are a few concepts that can shorten the setup time for fixtures:

- Hard stops can be used to position the part or assembly. These can include centering guides, dowel pins, guide blocks, etc.
- Templates or master parts or subassemblies will allow any required adjustments to be made quickly by securing adjusting devices against the template or master.

- U cuts, V cuts, wedges, or pear-shaped holes can be used to locate the part or subassembly quickly in the fixture.

- C washers (cutout washers) will allow for quick washer installation where washers are used to locate the part.

- Ratchet or power tools will accelerate any required threaded fastening steps.

- Standard clamps on fixtures will rapidly secure parts in the fixture. Over-center clamps such as toggles or camlocks serve this purpose well, because they secure parts with a single motion. Air-driven automatic clamps can be even quicker.

- Fixture designs that allow for installation of the part or subassembly in only one orientation will eliminate any confusion or incorrect installation during setup.

- Retractable bolts and quick-release fasteners can eliminate the need for screwdrivers, pliers, wrenches, or other hand tools.

- Magnets can be used to assist in pulling the part or subassembly into the fixture.

Figure 17.2 Fixturing for a CNC mill. This fixture has quick release fasteners and component positioning locators to accelerate the setup. This work center has multiple fixtures like this one, which allow the setups to occur offline to not interfere with run time on the mill. The loaded fixtures are moved to the mill's working area by an automated system.

Cutting Tools

Cutting tools used on lathes, drills, and milling machines can be designed for quick installation, with hard stops to eliminate the need for making adjustments or other compensations. Magnetic or air-assisted chucks can accelerate installing and securing the tool on the machine. Cutting tools can be preloaded into chucks, speeding installation into the machining center. Storing cutting tools in racks located on the machines where they are used shortens retrieval times. Some machines automatically install cutting tools, which greatly reduces that portion of the setup.

Figure 17.3 Preloaded drills. These chucks are preloaded with required drills and stored in organized containers near the lathe. This offline setup allows quick installation on the lathe, minimizing interference with the lathe's runtime.

Injection Molding and Die Casting

Setup times for injection molding and die casting equipment can be relatively long due to the tool weight, tool storage location, hose connections, and the requirement to run many parts to bring the equipment up to operating temperature. Parts made during this temperature adjustment phase will be nonconforming because the equipment has not yet reached operating temperature (these parts will ultimately be recast, remelted, or scrapped). Multiple hose connections may be required to inject the material to be molded or cast into the dies, and for the coolant needed to keep the tool temperatures in an acceptable range. The dies are often so heavy that forklift trucks are needed for removal and transport. The setup process in an injection molding or die casting operation can literally take hours and sometimes days.

A biomedical equipment manufacturer that used injection molding attacked the setup challenge aggressively and attained significant cost reductions. This organization designed an overhead rail system for its molds to eliminate the need for forklift trucks and manual lifting. They stored their molds in a heated storage area so that the molds were kept near their operating temperature (this eliminated the need to make hundreds of nonconforming parts to bring the molds up to temperature). They incorporated quick-release hoses to reduce the time to connect and tighten the feed and coolant tubes. This organization reduced injection molding setup times from several hours to less than fifteen minutes. Their payback was huge.

Storage Locations

Much time is lost during setups because of the need to retrieve material, tools, fixtures, drawings, and other items. Locating work-center-specific inventory racks, fixture racks, tool cabinets, shadowboards, tooling, drawings, and other required items near where the setup is performed can reduce this portion of the setup. The intent, as mentioned earlier in the lost time discussion, is to reduce time lost due to the operator walking or otherwise not being in the work center.

Figure 17.4 Cutting tool storage. These drills are preloaded into chucks and stored in an organized manner near where they are needed.

Pre-Kitting

The concept of pre-kitting involves packaging all items needed for a setup offline and then delivering the kit to the work center before the setup is needed. This makes everything needed instantly available. The kit can include tools, fixtures, gages, and any other equipment needed for the setup.

Procedures

Documentation outlining required setup steps, tools, and so on, can reduce times by eliminating confusion related to the setup. Digitally illustrated work instructions can help to shorten the process.

Moving Setups Offline

Inline setups are those that are performed on the production equipment. As such, inline setups prevent using the production equipment during the setup period. Offline setups take the setup off the production equipment. The concept here is to minimize the time the machine or other production equipment is not running because of the setup interfering with the machine's availability to make product. Ideally (and this is especially true on automated equipment), the same operator who operates the production equipment can work on the next setup while the prior job is running. Approaches for accomplishing this include:

- Fabricating multiple fixtures that allow setting up the next job while the prior job is running (see Figure 17.2).
- Heating the injection molding dies as described earlier for the biomedical equipment manufacturer example.
- Loading paint containers with the next color while the prior job is running.

The goal the setup reduction team should strive for is to move all setup work into an offline setup mode to maximize available runtime.

Who Should Do This Work

The setup time reduction team should include the operators who actually do the setups, the operators who make the parts (they may not be the same people), the work center supervisor, a manufacturing engineer, a tool designer assigned to the area, and a CNC programmer if the parts are made on a CNC machine.

We can make an argument for using the people assigned to the work center for this team, and we can also make an argument for assigning competent, knowledgeable people with the same skill set who are not assigned to the work center. The people already assigned to the work center may not be able to view the setup reduction challenge with a fresh perspective; people from outside the work center will not have the same product knowledge as do people actually performing the setup and making the parts. There is no textbook answer to this, other than to recognize these trade-offs and to assign people accordingly. Good setup reduction team leadership is probably the most important factor in successfully designing and implementing shortened setups.

Risks

The only risk in reducing setup times is that if the resulting setup approach is inadequate, nonconforming material may result. Normal precautions (such as inspecting the first parts produced using the shortened setup) suitably mitigate this risk.

References

J.P. Tanner, *Manufacturing Engineering*, New York, Marcel Dekker, 1991.

V. Chiles, and P. Williams, *Streamlined Manufacture: The Art of Setup Time Reduction*, Burlington, Butterworth Heinemann, 2000.

S. Kalpakjian, *Manufacturing Engineering and Technology*, Boston, Addison Wesley Publishing Company, 1992.

18

Material-Handling Improvements

The Bottom Line

Material-handling improvements can reduce costs by reducing how often and how far material has to be transported, and by eliminating the potential for handling damage or contamination during movement and storage.

Key Questions

Do we have any issues with material-handling damage?
Are we doing anything to reduce how often and how far we have to move material?

The Material-Handling Improvement Road Map

Material handling refers to activities and equipment associated with material movement from one point to another. Material-handling improvements can reduce cost in areas related to:

- Transport and handling reduction.
- Storage location elimination.

- Material-handling damage.
- Contamination elimination.
- Improved material handling equipment.

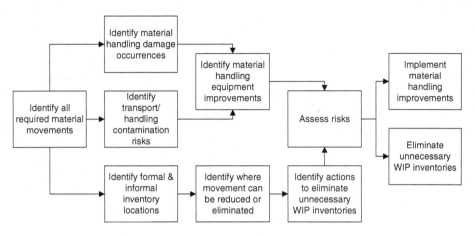

Figure 18.1 Recommended material-handling improvement approach.

Transport and Handling Reduction

In an earlier discussion related to lost time, we covered the importance of locating tools, material, and other equipment near where these items are actually needed. This eliminates lost time and drives cost down. Techniques for achieving this include local inventories of required items and tools, sequencing work centers in a linear manner, and minimizing the distances between work centers.

Minimizing the distances between work centers usually means locating the work centers right next to each other, but as described earlier, this is not always practical. Sometimes walls interfere with transferring material from work center to work center. Where this occurs, it may be possible to create doors or windows that allow moving material through the wall rather than having to leave the room, walk around the wall, and enter the next room.

In the chapter on work-flow optimization, we discussed monuments (i.e., production machines and other equipment that cannot be easily relocated). When monuments are encountered, it may be necessary to move material longer distances, but there are options:

- One option is to relocate the entire process closer to the monument.
- Another option is to examine the work performed by the monument and determining if the work can be done closer to where it is needed. For example, if a

hole-drilling operation is performed on a large machine that cannot be easily relocated, perhaps a small drill press in the work center will suffice.

- A third option is modifying the product design to eliminate the features necessitating the move to the monument.

In general, anything that can be done to minimize distances between work centers is a good thing. From material handling damage and contamination perspectives, it generally means the item will be handled and exposed less.

Storage Location Elimination

When it makes sense to do so, the organization should eliminate unnecessary inventory storage locations. There are two situations to consider:

- In one situation, the organization stores inventory in the work centers where it is needed (e.g., fasteners, wire, etc.). This is okay, because it is being done to reduce lost time.
- In the other situation, work-in-process (WIP) inventory storage locations emerge because the process is unbalanced. WIP inventory accumulates in these interim locations because the work centers complete work at different rates. These are the inventory locations we want to eliminate.

Whenever possible, it is a good idea to eliminate interim WIP inventory storage locations of the type described in the previous list. The ideal situation is that the process flows in a balanced manner, never accumulating inventory along the way due to rate differences from one work center to the next. As these WIP storage locations are eliminated, the need to handle the inventory is reduced. This reduces cost for three reasons:

- Less labor is required.
- The inventory decreases.
- There is less opportunity for damaging or contaminating the inventory in storage or during handling.

Transport Packaging Improvement

Material-handling damage occurs when parts and assemblies bang into each other, bang into other things, fall, or are subjected to adverse environments. Addressing this problem requires identifying where the damage is occurring, why it is occurring, and eliminating the causes. Once the dominant material-handling damage occurrences are identified, identifying the cause is straightforward. Material-handling damage usually occurs because the parts are moved without

Figure 18.2 Preventing material-handling damage. The small machined parts on the left are transported in compressed fiber (cardboard) trays. The larger cylinders are packaged individually. Both approaches prevent the parts from damaging each other during transport, and from being damaged by other objects.

adequate protection, or the people moving the material have not been trained properly. Correcting material-handling damage usually involves improving transport fixtures, training the material handlers, incorporating improved packaging, incorporating improved protection while the material is stored or transported (see Figure 18.2), and minimizing movement and handling.

The quality assurance department should be able to help by rapidly identifying the parts and subassemblies experiencing material-handling damage most frequently. If this information is not available, it is usually a simple matter to collect it either by reviewing existing records or by asking the people who see the damage. Just looking at areas where material is stored will reveal what is occurring.

Contamination

Sometimes contamination occurs while items are being transported. Fixtures or carts used to transport parts, subassemblies, or materials can become contaminated and transfer the contaminants to their cargo. If the fixtures do not adequately protect their cargo from contaminants, the parts can become contaminated as they are transported through a contaminated area (e.g., moving electronic or optical assemblies through work areas that use silicone sprays). Sometimes parts or materials become contaminated if they are transported from one building to the next and they are exposed to rain, airborne dust, or other outside contaminants.

A review of the organization's nonconformances and discussions with the people who administer rejected items will reveal if material-handling-related contamination is a problem. If it is, it can be controlled by reducing material movement and improving protection from contamination during movement.

Improved Material-Handling Equipment

Material-handling equipment improvements can eliminate damage, shorten setup times, eliminate the need for a person to physically transport the item, and eliminate contamination. Here are suggested approaches:

- In some cases, it may be possible to combine transport containers and fixtures. If a part is transported in a container and then subsequently loaded into a fixture, designing a container that both houses the item for transport and secures it in position for the next manufacturing operation is a great approach. It eliminates taking the item in and out of the container, and it eliminates the need for a separate fixture. This concept is particularly well suited for assembly operations requiring fixtures and transport containers. The cost of designing and fabricating such a combined container-fixture needs to be compared to the cost of separate containers and fixtures, but usually the organization can realize cost reductions using with this approach.
- Containers can sometimes be designed such that they are used by both the customer and the supplier. Sometimes they can be recycled, further reducing cost.
- Conveyors can be used to reduce material-handling labor.
- Padded carts and containers can be used to prevent damage to the item being transported.
- Rails can be incorporated to assure that transport carts are not allowed to bang into components.
- Cartons can incorporate dividers to prevent items being transported from damaging each other (see Figure 18.2).

Who Should Do This Work

Implementing the topics addressed in this chapter will require help from several functions. The cost-reduction team can take a lead role in the effort to improve material handling, with support from the material, engineering, manufacturing, purchasing, and quality assurance departments.

Risks

Risks are minimal, and it is relatively easy to realize quick cost reductions in this area. There is a risk of disrupting production as improvement actions are implemented, particularly when eliminating unnecessary inventory storage locations. This risk can be minimized by balancing the work centers' rates before eliminating the unnecessary inventory locations.

References

S. Kalpakjian, *Manufacturing Engineering and Technology*, Boston, Addison Wesley Publishing Company, 1992.

J.P. Tanner, *Manufacturing Engineering*, New York, Marcel Dekker, 1991.

J.H. Berk, *Quality Management for the Technology Sector*, Burlington, Elsevier, 2000.

19

Scrap and Rework Reduction

The Bottom Line

Scrap and rework are often significant, and the organization's cost-reduction effort should focus on this area. Documenting all nonconformances and then sorting the data will identify dominant nonconformances. Teams should be assigned to correct the causes of these dominant nonconformances. Doing so requires identifying all potential nonconformance causes, objectively evaluating each, and implementing appropriate corrective actions to prevent recurrence. This activity can be extended to incorporate corrective actions for likely potential causes in addition to confirmed causes, which will help to prevent future nonconformances. The organization should move from a detection-oriented quality management approach to an approach that focuses on preventing defects.

Key Questions

What are our most frequently occurring nonconformances?

What are our most expensive nonconformances?

What is our approach for eliminating recurring nonconformances?

Do we measure our progress in eliminating recurring nonconformances?

The Scrap and Rework Reduction Road Map

This chapter addresses how to reduce scrap and rework. This is an incredibly rich cost-reduction opportunity.

In most organizations, when a manufacturer produces a product it is inspected at various stages during the production cycle. In some cases, the product is found to be nonconforming. *Nonconforming* means the product does not meet its drawing or specification requirements. When this occurs, the organization needs to decide what to do with the nonconforming item. The manufacturer can:

- Use it "as is" in the nonconforming condition if the product is deemed acceptable for use in its nonconforming condition and it is acceptable to the customer.

- Rework the nonconforming material such that it meets the drawing requirements.

- Repair the nonconforming material in a manner that doesn't exactly meet the drawing requirements, but the repair makes it suitable for use (again, if this is suitable to the customer).

- Scrap the nonconforming item if it cannot be reworked or repaired economically.

The cost of these activities is high. Most people are surprised when they learn how high the costs of poor quality are. Informed estimates consistently place the cost of poor quality between 20 and 50 percent of the total cost of goods sold. That is a huge number, and it represents a huge opportunity. If the organization can reduce the amount of nonconforming material it produces, the

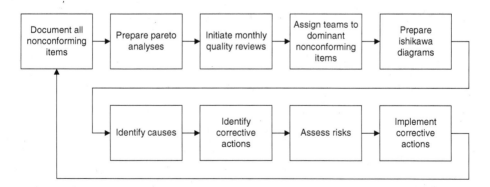

Figure 19.1 Recommended scrap and rework reduction approach.

cost of reworking, repairing, remaking, and administering nonconforming material will lower costs significantly.

The Hidden Factory

The *hidden factory* is a term used to describe the work that goes on to address nonconforming material. This includes the rework, repair, and remaking of nonconforming material, as well as the work required to administer the nonconforming material. The area where this work actually occurs is usually not actually hidden.[1] Usually the organization views it as a normal part of the manufacturing process. It is such an ingrained part of the process that management doesn't view it as unusual or unnecessary, but it is.

The hidden factory is where an organization pays twice for the same thing. The organization pays for the work the first time it is performed. Then they pay for it again when it is reworked, repaired, or scrapped and remade. Sometimes the rework or repair activities have benign names that sound as if they should be a normal part of the process (such as "touch up" or "pre-delivery inspection"), but make no mistake: this is all work that could be eliminated if the product didn't need to be touched up or inspected again.

An Example: The Hidden Factory in a Motor Home Company

We considered a motor home manufacturer in an earlier example, and we will return to that example here. It had 200 employees on its assembly line and in the supporting work centers. Here's how their process worked:

- As the motor homes progressed through stations on the assembly line, inspectors examined the work at each station and noted the nonconformances. They did this by attaching strips of red tape on the coach to designate nonconforming areas.

- The supervisors in each workstation reviewed the nonconformances and assigned operators to rework the rejected areas. This typically was accomplished by having operators from the work center move with the motor home to the next work center to correct the deficiencies. As the rework was completed, the person doing the rework removed the red tape strips.

- At the end of the assembly line, after the motor homes were considered complete, another set of inspectors subjected the motor home to a "final"

---

1. Sometimes, however, it is actually hidden, as occurs in the case when an item is found to be nonconforming and it is furtively scrapped without the event being documented.

inspection. These inspectors noted between 100 and 200 defects on each motor home. The defects were again identified with strips of red tape.

- Approximately fifty additional full-time workers corrected the defects found in final inspection. The folks who did this work were assigned to the touchup crew, which was another normal part of the process. They also removed the strips of red tape as they completed the rework.

- After the touchup crew completed their work, the company's president and the vice president of marketing would randomly select a motor home and personally inspect it. They always found between twenty and fifty rejectable conditions, which they also identified with red tape. These two people also viewed this activity as proof of their commitment to quality. They had been doing this for years.

- The touchup crew returned to correct the deficiencies found by the president and the vice president (who were unquestionably two of the highest paid quality assurance inspectors who ever lived).

If we take a step back and consider what was going on in this company, we can recognize that the amount of inspection and rework was huge. The touchup crew alone was 20 percent of the workforce. Think about that for a moment. Their work alone, all of which was to correct others' errors, constituted 20 percent of all labor.

The in-process inspection and rework further added to the cost of poor quality, and it was considered a normal part of the process. It was obvious and in plain sight, but no one saw it as unusual because they had been conditioned to believe it was the way manufacturing was supposed to work. It was a hidden factory within a factory, yet it was in plain sight.

If anyone had told the president of this company that he could eliminate 20 percent of the manufacturing costs, he would have jumped at the chance to do so. Yet there it was, right in front of everyone, and no one saw it for what it was.

Prevention Versus Detection

So, how can this be done?

To significantly reduce the costs of poor quality, an organization's approach to manufacturing has to change. In the motor home company described earlier (and in many, many manufacturing organizations), the quality approach is based on detecting defects and then correcting them. To significantly reduce defects, an organization needs to change its thinking. Instead of detecting and correcting defects, the focus has to be on *preventing* defects. When nonconformances are prevented, there's no need to rework or scrap them.

Shifting from a detection-oriented to a prevention-oriented approach consists of the following steps:

- The organization has to document all nonconformances. It should already be doing this; if it is not, it needs to start. Some will argue that small nonconformances that are easily corrected don't have to be documented. Don't listen to them. Document all nonconformances. Some will argue that it will become an administrative burden because of all the nonconformances. Don't listen to them, either. When the number of nonconformances gets smaller (and it will), documenting them won't be such a big deal.

- The nonconformance documentation has to be collected and categorized by part number, work center, and defect type. The quality assurance group should already be doing this; if not, they need to start.

- These data need to be displayed in monthly Pareto charts that show the frequency of occurrence versus defect type, part number, or work center (the organization can decide how it wants to sort the data). Similar Pareto charts should be prepared showing cost[2] versus defect type, part number, or work center. Figures 19.2 and 19.3 show typical Pareto charts. The concept is that the organization will be able to see which defects are occurring most frequently and which defects are costing the most. This is important, because it makes more sense to attack the dominant items first. The cost reductions will come more quickly and be more significant.

Problem Solving

Once the organization has prepared the data, it should form teams to identify the nonconformance causes and assign corrective actions. The team should include operators from the work center, the work center supervisor, an engineer, and others (depending on the nature of the nonconformance). It's important to include a mix of people. In many cases, the operators (the people who are closest to and actually do the work) will have the best insights into nonconformance causes.

The four-step problem-solving process shown in Figure 19.4 outlines an approach that works well.

---

2. When calculating costs for each nonconformance, the cost is substantially more than just the cost of the rejected item. It should include the cost of all labor that went into the part up to that point and the costs associated with administering and dispositioning the nonconformance. The organization's accounting group can often assist here.

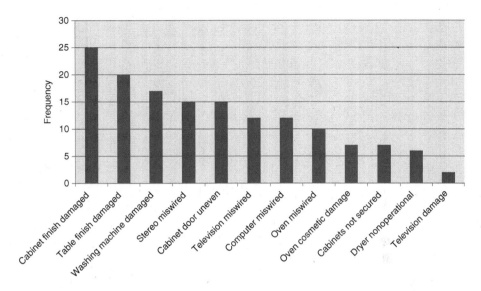

Figure 19.2 Motor coach interior July nonconformances. This Pareto chart shows the most-frequently occurring interior nonconformances for the month of July. The chart suggests that preventing finish damage before and during installation will eliminate a large percentage of the nonconformances.

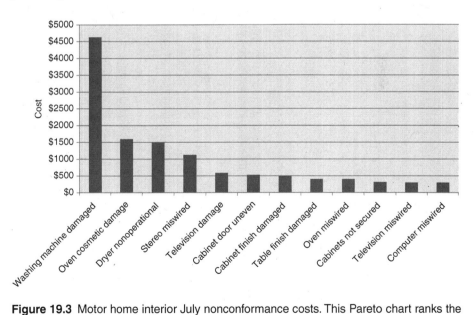

Figure 19.3 Motor home interior July nonconformance costs. This Pareto chart ranks the nonconformances by cost. Washing machine damage is the largest cost. The organization should focus on finding and eliminating the causes of the damaged washing machines.

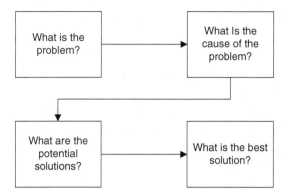

Figure 19.4 The four-step problem-solving process.

The first step, defining the problem, is the most difficult. The group should spend some time making sure all agree on the problem. Some will define potential causes and list these as the problem, but doing so often precludes listing other potential causes. In most cases, the nonconformance description is actually a good statement of the problem.

The second step requires identifying the cause of the problem. The organization can use any of several techniques here. These include brainstorming, mind-mapping, Ishikawa diagrams, and fault tree analysis. Brainstorming and Ishikawa diagrams (see Figure 19.5) generally work well.

Another Example: Ocean Racing Yacht Composite Sail Failures

A high-performance racing yacht composite sail manufacturer experienced nonconformances both in-house during production and in service during ocean races. This was a serious problem, because sail failures during a race could result in loss of the boat and its crew.

The process consisted of the following steps:

- Applying melted adhesive to thin plastic film.
- Positioning the film on full-size forms that duplicated the sail's shape when it was in use.
- Applying additional adhesive to the film on the form.
- Laying another film on top of the adhesive.
- Heating the adhesive with a heat blanket to melt it, which resulted in the layers bonding together.

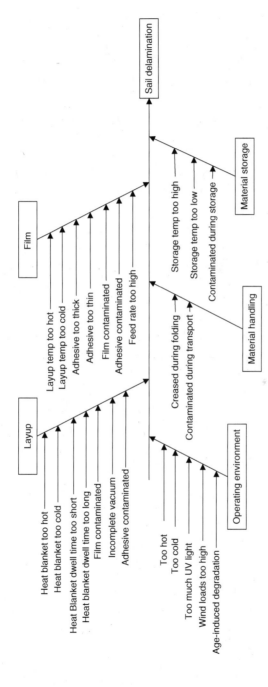

Figure 19.5 Ocean racing yacht sail delamination Ishikawa diagram. This manufacturer produced composite sails for high-performance ocean racing yachts. When several sails delaminated in use, the manufacturer used an Ishikawa diagram to assist in developing a list of potential causes. The box at the head of the fish represents the problem statement and the boxes forming the skeleton are potential causes.

The nonconformances occurred when the completed sail's layers delaminated. The manufacturer prepared an Ishikawa diagram (Figure 19.5) to identify potential nonconformance causes. After identifying the potential causes, the company then evaluated each. To its surprise, it found many potential causes, each capable of inducing sail delimination (they had previously assumed all of the failures had been induced by the same cause). Initial investigation showed that although the hypothesized potential causes had not actually caused any of the failures under investigation, they could do so in the future. The company opted to implement preventive actions to preclude these potential causes from occurring. The company continued to develop the Ishikawa diagram, ultimately identifying the potential cause that actually induced the failure.

Evaluating Potential Causes

After identifying potential nonconformance causes, the team should objectively evaluate each. Sometimes causes are dismissed too quickly because of personalities or group dynamics. It's important to objectively assess all potential causes, and not simply jump to the one favored by the loudest or most forceful group members.

This objective evaluation will frequently involve many activities and many other people in the organization. It may be necessary for the purchasing group and the engineers to work with suppliers if potential causes exist in supplier materials. The team should study the process to determine if all steps make sense and are being performed correctly. It may be necessary to perform special tests to evaluate potential causes. Many other actions will probably be required, depending on the nature of the hypothesized potential causes. The assessment of these potential causes must be objective and it must be thorough.

Corrective Action

After the team has completed the aforementioned activities and converged on the actual nonconformance cause, the next step is to evaluate potential corrective actions. Based on the team's evaluation, these corrective actions may include:

- Process changes (e.g., a revised work sequence, the use of different tools, different machine feeds and speeds, improved lighting, etc.).
- Design changes (e.g., features to prevent incorrect installation orientation, different materials, eliminating stress raisers, etc.).
- Requirements relaxation (e.g., increasing drawing tolerances that allow the nonconforming condition).
- Training (to give the operators better knowledge or skills).

Generally, the best corrective actions are those that make it impossible for the nonconformance to recur, or that greatly reduce the likelihood of the nonconformance recurring. Design, process, and requirements changes generally fall into this category.

Less reliable corrective actions focus on having people not make mistakes (these are less reliable because people will make mistakes). These corrective actions may be the only ones available in the short term, but if so, longer-term corrective actions (such as design or process changes) should be considered for subsequent implementation.

During the course of its investigation, the team assigned to a specific nonconformance will most likely identify several potential nonconformance causes, and then narrow their list down to a specific cause. Here's a question: Should the team implement corrective actions for only the specific cause known to have induced the nonconformance, or should it also implement corrective actions for other potential causes?

Consider the composite sail manufacturer described earlier, and how they handled this question. One of the significant advantages associated with objectively identifying all potential causes of a problem is that it can be used to drive *preventive* actions as discussed earlier in this chapter. The team should not be restricted to only fixing confirmed causes that have already occurred. If other actions can be implemented to prevent potential nonconformance causes, these should also be implemented if it is cost-effective to do so.

Managing the Effort

There are several actions required to manage an effort to reduce costs through quality improvement:

- The organization should have monthly quality reviews focused on assessing the Pareto charts described earlier. These reviews should also assess the status of efforts to improve specific quality deficiencies. These monthly reviews are important, and we always recommend that the president or general manager attend. This is another situation in which interest and support from the top will drive the effort's success.

- As dominant nonconformances are revealed through Pareto analysis, the organization should assign multidisciplinary teams to eliminate the nonconformance using the techniques outlined in this chapter.

- The teams assigned to eliminating each dominant nonconformance should report their progress at these meetings.

- The teams assigned to eliminate each nonconformance should each prepare an action plan outlining the potential causes, what actions are required to evaluate the potential causes, who is assigned to each action, the date by which the action should be complete, and the status of each action.

The approach described in this chapter will take the organization from a detection-oriented management philosophy to one that is preventive in nature. By using the approach described here, the quality improvement and the cost reductions will be significant. It's not unusual for organizations committed to scrap and rework reduction to realize their goals within three to six months.

Who Should Do This Work

Senior management has to be actively involved in overseeing the transition from a detection-oriented to a prevention-oriented manufacturing approach. The magnitude of the cost reductions deserves senior management attention. This involvement should include the actions described in this chapter as well as others. Sometimes just taking a walk behind the plant and looking into the scrap bins can be enlightening.

The quality assurance group should take the lead in documenting and summarizing nonconformances. The finance group should assist this effort by providing cost data for nonconformances at the point of rejection.

The organization's management should select the nonconformances targeted for elimination based on what the Pareto analyses show.

The teams assigned to eliminate nonconformances should include members from the manufacturing, purchasing, engineering, quality assurance, and other groups based on the nature of the nonconformance. Having a multidisciplinary team work these issues generally assures a more objective analysis.

Risks

The biggest risk in this area is management failing to recognize the enormous cost reductions associated with scrap and rework reduction, and in moving to a prevention-oriented manufacturing approach. Think about that motor home company president (and his marketing vice president) who randomly selected one coach for their inspection every day. They found numerous defects after the coach had been repeatedly inspected, and they continued doing this every day for years. What kind of quality did they think the other coaches had? They were proving, without intending to so do, that detection-oriented systems just don't work.

Lack of management support is another risk. If the people at the top are committed to making scrap and rework causes go away, the rest of the organization will follow along. A good way to mitigate this risk is to show some early successes, and quantify the associated savings.

Weak problem-solving skills exist in some organizations. Where this is the case, problem-solving and failure-analysis training can help enormously.

Failing to identify and objectively evaluate all potential nonconformance causes is a big risk. As mentioned earlier, sometimes egos get in the way of solving problems, or people don't want to admit their area caused the problem. Using multidisciplinary problem-solving teams will mitigate this risk.

References

J.H. Berk, *Quality Management for the Technology Sector*, Burlington, Elsevier, 2000.

J.H. Berk, *Systems Failure Analysis*, Materials Park, ASM International, 2009.

20

Work Center Cleanliness

The Bottom Line

Keeping the plant clean and well organized reduces cost by reducing inventory, making needed items easier to find, improving morale, and increasing efficiency. The 5S and other cleanliness programs can help get a plant organized and sustain a clean work environment. Management can assist by participating in weekly plant tours to assess the plant's condition.

Key Questions

Are we satisfied with the cleanliness of our work environment, both in the shop and in the office?

How often does management get into the plant?

What is our approach for keeping our facility clean and organized?

The Cleanliness Road Map

Clean work areas run more efficiently. Henry Ford (who introduced mass production in the automobile industry) and Frederick Taylor (the father of modern industrial engineering) realized this a century ago. Most people intuitively

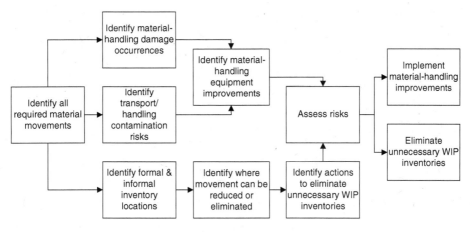

Figure 20.1 Recommended work area cleanliness approach.

grasp it. There are many reasons a clean shop runs better, including the ability to locate needed items quickly, the elimination of unneeded materials, and the natural productivity boost that occurs when working in a neat and organized environment.

Knowing that a clean work area runs more efficiently and actually attaining a clean factory are two separate things, however. Many managers anguish over how to get the workforce to clean up the plant and keep it that way. Many cleanup efforts are short-lived, and as the plant becomes more cluttered, its operating costs rise.

This chapter presents two proven approaches for running a clean shop: the 5S system and a competitive cleanliness program.

The 5S Program

The term *5S* refers to a management approach centered on five words, all of which begin with the letter S. The 5S approach was originally christened in Japan, but folks here in America managed to find five English translations that also start with S. These are:

- Sort (the Japanese word is Seiri)
- Stabilize (the Japanese word is Seiton)
- Shine (the Japanese word is Seiso)
- Standardize (the Japanese word is Seiketsu)
- Sustain (the Japanese word is Shitsuke).

The program is generally taught as one of Japanese origin, but the concepts in it resulted from the Japanese studying and adopting the work of Henry Ford and Frederick Taylor.

The 5S program simplifies an approach for cleaning and organizing a shop, and keeping it that way. Its precepts are nothing more than common sense, but the 5S program's organization and simplicity make it a good way to communicate the need for a clean and orderly shop. Here's an explanation of each "S" in the 5S program:

- Sorting means assessing and separating everything in the work center into either essential or nonessential items. Only essential items should be kept in the work center. Other nonessential items should be removed. They can be stored, sold, or scrapped, depending on how often they might be needed and their value. The emphasis is on not keeping things around just because they might be needed someday. The idea is that this frees up workspace and makes it easier to find essential items when they are needed.

- Stabilizing means arranging what is left in the work center in an organized manner. Sometimes this phase of the 5S program is called straightening, systematically organizing, or setting in order. The concept is that tools and other work center things are placed where needed such that their locations promote efficiency. The locations should be identified or labeled, so that work center items are always kept in the same spot. Shadow boards, as shown in Figure 20.2, are frequently used for this purpose.

- Shine refers to keeping the work center clean. Other S-based descriptors for this phase are sweeping and systematic cleaning. The emphasis here is that this is an ongoing theme and that cleaning should occur daily. Work centers should not need to be cleaned or tidied for special occasions; they should "shine" all the time. There are several advantages to this. In addition to maintaining a culture of cleanliness, clean work centers make it easy to see leaking fluids, equipment problems, missing items, etc., and the problems can be addressed before they become showstoppers.

- Standardize is the fourth phase of the 5S program, and it means standardizing work methods so that they minimize process variability. The idea is that work is broken down into its simplest elements (sometimes this phase also referred to as simplifying), and operators are assigned tasks that can be performed in a similar manner each time. This approach is directly based on Frederick Taylor's scientific management principles.

- Sustain is the last S. This means maintaining practices, actions, and methods developed in the previous S categories; the concept here is that these practices are sustained. An associated precept is that a 5S program should have

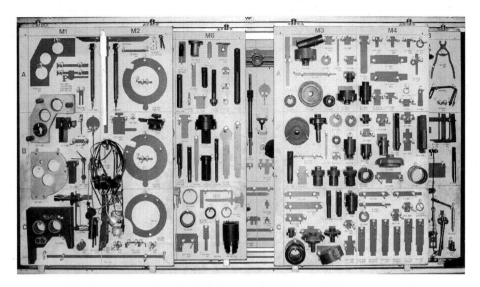

Figure 20.2 Tool shadow board. Shadow boards outline tool shapes on a board, which makes it easy to store and find tools.

an audit function to make sure that the previous activities, once implemented, continue to guide how the organization operates.

Some organizations have a 6S program, with safety constituting the sixth S. 5S adherents have pointed out that safety naturally results from a 5S program, as a clean and organized workplace creates a safe environment. Whether the program is denoted as a 5S or a 6S program is a detail that is probably not worth debating.

The 5S program works where there is sustained management interest. A weekly management walk around the shop will foster a clean environment if management requires that the facility be kept clean and orderly. In some organizations, senior management rarely ventures into the shop, and that's a poor way to run a company.

Sometimes organizations using the 5S program rely on posters and signs emphasizing the approach. The effect of motivational posters on human behavior is questionable. Management interest combined with a willingness to explore the factory and ask questions can have a profound impact.

The Competitive Cleanliness Approach

We first observed a competitive cleanliness approach at a Baxter facility in Mayaguez, Puerto Rico. This facility manufactured cardiovascular surgical kits, so the need for keeping the shop clean was extremely important.

The competitive cleanliness approach is simple, but it works. We've implemented it in several plants and the response was impressive.

Cleanliness program										
Work center	3-Jul	10-Jul	17-Jul	24-Jul	31-Jul	1-Aug	8-Aug	15-Aug	21-Aug	28-Aug
Receiving	3	3	4	4	5					
Receiving inspection	4	4	4	4	5					
Machining	4	3	5	5	5					
Paint preparation	3	3	3	3	4					
Paint	5	5	5	5	5					
Assembly	2	3	3	4	4					
Test	4	4	5	5	5					
Final inspection	5	5	4	5	5					
Shipping	4	5	5	5	5					
Warehouse	3	4	5	5	5					
Total plant	37	39	43	45	48					

Figure 20.3 A competitive cleanliness scoreboard. This board should be posted in a prominent location in the plant. The work centers receive a weekly cleanliness score, and the total plant score is shown on the bottom.

The idea is that the company posts a large scoreboard in a prominent area, with the simple title "Cleanliness Program" posted at the top of the board. The scoreboard contains a matrix, with the leftmost column denoting work centers and the top row denoting weekly inspection dates. Management then tours the facility on a weekly basis and assigns a 1 to 5 rating for cleanliness to each work center, with 5 representing the best possible rating. The scores can be summed at the bottom for an overall facility rating.

This simple program works. In one manufacturing facility that struggled with shop cleanliness for years, the company's management adopted the program. Due to an oversight, management didn't explain to the employees what they were doing. The company's leaders simply erected the scoreboard and began the weekly work center reviews without explaining anything. Employees immediately understood what the program was all about, and by the third week, the facility was dramatically cleaner and more organized. The company tracked labor hours per assembly as its key performance metric, and that improved as the shop became cleaner. The program's success, without any motivational posters, training, or other exhortations, was impressive. The element of work center competition is probably one of the key factors underlying this shop cleanliness program's success.

Why it Works

Whether the organization adopts a formalized program (such as a competitive cleanliness program or the 5S program), or it simply has interested senior management tour the facility and ask questions, the facility will become cleaner and more organized. Cleaner and more organized facilities run more efficiently. As mentioned at the start of this chapter, some of this is due to being able to find things quickly, some of it is due to eliminating excess inventory, and some of it is probably psychological. When work centers are clean and free of unnecessary items, it's easier to remain focused.

The concept of a clean and organized work environment is not restricted only to the factory. Many organizations find that requiring office workers to keep their work areas organized results in similar efficiency and cost improvements.

Who Should Do This Work

The cost-reduction team, senior management, or the manufacturing organization can initiate a program to organize and keep the workplace clean. Sustained senior management interest will sustain the program. All employees in the manufacturing and office areas can contribute to keeping the workplace clean.

Risks

There really are no risks or other downsides associated with maintaining a clean work environment. Some folks might express a concern about losing production time for cleaning and organizing activities, but the small amount of time this requires once the workplace has been cleaned will be more than offset by efficiency gains.

References

H. Hirano, *Five Pillars of the Visual Workplace*, New York, Productivity Press, 1995.
B. Meakin, *Model Factories, and Villages: Ideal Conditions of Labor and Housing*, Ithaca, Cornell University, 2009.

Part IV

Design

The design of a product and its packaging strongly influence cost. Opportunities for taking cost out of the design can be found in the following areas:

- Design approach.
- Requirements relaxation.
- Tolerance relaxtion.
- Materials substitution.
- Packaging approach.

21

The Design Approach

The Bottom Line

The design process should include defining the requirements, brainstorming alternative design approaches, and then selecting the approach that best meets the requirements. One of the requirements should be minimizing cost. This is true for both new product developments and redesigns of existing products. Redesigning existing products can reduce cost if cost reduction was not a significant factor during development or if new technologies offer less costly design approaches.

Key Questions

How do we select design concepts when developing new products?

How do we consider cost during the design process?

Are we doing anything to take cost out of our existing product designs?

How old are our existing designs?

Have new technologies emerged that might lower our existing costs?

Do we have control over our product's design, or do we need permission from the customer to modify it?

The Design Approach Road Map

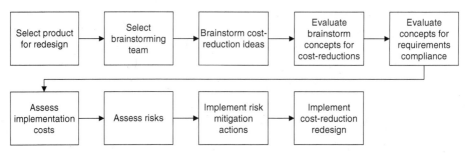

Figure 21.1 Recommended design cost minimization approach.

The Design Process

Engineers design products in response to a set of requirements by identifying and evaluating all approaches for meeting the requirements, and then selecting the best approach. Figure 21.2 shows the process.

Identifying the requirements is the first step. This may include inputs from the customer, the sales organization, the engineering group, and others. Identifying the requirements should also include internal requirements, such as the processes available for manufacturing the product, how much the product should cost, the desired profit margin, etc. The next chapter will focus on evaluating requirements to determine if any can be relaxed to realize cost reductions, so we'll bypass this area for now.

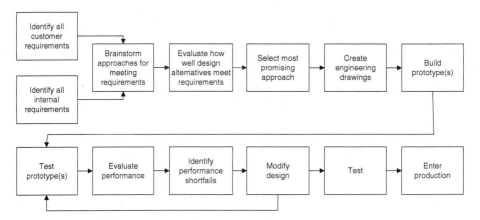

Figure 21.2 The design process. After identifying all requirements, engineers identify design alternatives and then select the alternative that best satisfies all requirements.

The remaining steps involve brainstorming, evaluating, and selecting the best approach for meeting the requirements. That is our focus in this chapter. When developing new products, the design team should select the approach that best meets all requirements, including cost minimization. When evaluating redesigns of existing products, the challenge is to find an approach that still meets all requirements, but does so at a lower cost.

Lower cost alternatives for existing designs may exist for any of several reasons:

- The original design team may not have developed a complete alternatives list. They may have accepted the first workable concept without adequately identifying other alternatives.

- The original design team may not have had access to lower cost technologies because the technologies were not yet available.

- The original design team may not have been aware of alternative design approaches that would have lowered costs even though the technology existed. For example, airplanes used to use mechanical cables, pulleys, and levers to translate pilot inputs to movement of an airplane's flight control surfaces. In the 1970s (after aircraft had been flying for seventy years), General Dynamics did away with this bulky, expensive, and heavy approach when they designed the F-16. General Dynamics used electrical signals on the F-16 to control the flight control surface actuators. They advertised the F-16 as the world's first "fly-by-wire" aircraft, but it was not. The Germans used the approach during World War II on the Focke-Wolfe 190 fighter. Between World War II and the F-16 development program in the 1970s, aircraft designers did not consider the fly-by-wire concept because they did not know about it.

- The original design team may not have considered alternative manufacturing processes in the design. For example, an organization that uses machining as its primary forming method may have a mind-set that results in all of its designs using only machined parts. Castings, injection-molded plastic parts, or composite parts may be less expensive, but the design team might fail to consider these because they are conditioned to working with machined parts.

- The original design team may not have considered alternative lower cost materials. This might occur because the original team failed to consider these materials, or it may be because newer and less expensive materials were not available when the engineers developed the original design.

- The original design team may not have considered alternative lower cost fastening methods. This might occur for the same reasons (either the engineers missed the lower cost approach or new technology became available after the original development effort).

Brainstorming

Brainstorming is a powerful approach that works. It is easy to use and it is fun. The approach consists of defining an objective and then asking for inputs on potential ways to accomplish it. The task here is to either design a new product or redesign an existing product to reduce cost, and brainstorming works well when doing this. Here's a good way to plan and execute a product-design-related cost-reduction brainstorming session:

- Select bright people who are "out-of-the-box" thinkers.
- Notify the attendees a few days in advance of the brainstorming session and tell them what the topic will be. This will give them time to start thinking about the topic.
- Explain at the beginning of the brainstorming session that no one can criticize any of the ideas, and no judgments will be made on any of the ideas during the meeting. Let everyone know that the quantity of ideas is more important than the quality. The reason this is important is that if the quality of an idea is stressed, people will be more reluctant to offer ideas. Another reason is that even infeasible ideas tend to generate more ideas, and some of these might be breakthroughs.
- Have someone in the meeting capture the ideas on a large whiteboard or on an easel-mounted pad. This is better than recording the ideas on a small piece of paper or on a laptop where other people can't see them. When people in the meeting can see the ideas on a board or an easel, it will help them to think of more ideas.

At first, getting ideas to start flowing in most brainstorming sessions can be tough, but once the ideas start to flow, they will come in large waves. The challenge at that point will be capturing all of the ideas on the board or on the easel. After a while, the spontaneity and flood of ideas will slow. When this occurs, there are several ways to stimulate it again:

- Read the ideas already captured aloud. This may reignite the idea-generation process.
- Focus on each design concept one at a time and ask for similar concepts. Sometimes picking the keyword in each idea and asking for synonyms brings more ideas.
- Select the designs' quantitative characteristics and ask what would happen if these characteristics are increased or decreased. This may trigger new ideas.
- Ask if designs from completely unrelated products can be applied to the design being discussed in the brainstorming session. For example, the bolt-action rifle

was invented after an engineer saw a locking bolt on a gate. Many other stories exist where simple design concepts were applied to entirely new applications with great results.

- Pull on the ideas generated in the brainstorming session by asking the same questions a reporter might ask. These include the who, what, when, where, and why of proposed ideas. This may generate new ideas.

After the team has generated a list of potential cost-reduction ideas, the team members should evaluate each. Cost should be a consideration when this occurs on a new product, but as outlined at the beginning of this chapter, this does not always occur because cost may not have been a dominant factor during the original design effort.

If the team is doing this for an existing product as part of a cost-reduction effort, cost will naturally be one of the prime considerations. In this situation, all of the other design requirements must still be considered because the product will still have to meet the other requirements.[1]

Identifying each requirement during a new design (or a redesign) and finding the best alternative for satisfying all of the requirements is a normal part of this process. Each design situation is too unique to be explored here. Rather than attempting to do so, we will list several scenarios in which appropriate alternative design selections can result in lower cost.

- Many products can be designed or redesigned to use less costly materials. Some of the redesigns occur when new materials became available.when cost-reduction teams seek to make the products less expensive. Examples include using plastic instead of steel on such things as automobile interior components and bodies, using composite materials instead of aluminum on aircraft, etc. The use of alternative materials as a cost-reduction measure is covered in a later chapter.
- In many cases, adhesives can be substituted instead of threaded fasteners, rivets, or other mechanical fastening methods.
- Electrical or electronic controls can take the place of mechanical linkages in many applications. The F-16 fly-by-wire design mentioned earlier is a good example. Many other products (automobiles, manufacturing equipment, cameras, etc.) have switched from mechanical to electrical or electronic controls to reduce cost.

---

1. This assumes all of the other requirements are still valid. As mentioned earlier, the next chapter explores requirements relaxation as a cost-reduction measure.

- Holes and other shapes can be cast or molded into a product, rather than requiring secondary machining operations.
- Castings or molded parts can be used instead of machined parts.
- Computer-aided design optimization software can minimize the amount of material used while still meeting strength requirements.
- Using integrated circuit components instead of discrete components can lower costs.
- Wherever multiple fasteners are required, using fewer fasteners may still allow the product to meet its requirements. For example, if a cover is held on by eight screws, will four do the same job? If this can be done, costs will drop due to using fewer screws and washers and doing less machining and assembling.
- Using a single component to serve multiple functions can offer a cost savings. This is often done in aircraft, for example, where the same switch operates different systems when the aircraft is on the ground versus when it is airborne.

Who Should Do This Work

The cost-reduction team will need to rely on the engineering department's expertise in evaluating the impact of cost-reduction-driven redesigns on all of the product's requirements. Just asking the engineering department to take cost out of the product through a redesign may not always work well (see the risk discussion), so delegating this task in its entirety to the engineers is usually not a good idea. The best approach usually consists of the engineering department supporting the cost-reduction team by participating in the brainstorming sessions and by taking a lead role in evaluating any proposed redesigns' impact on requirements compliance.

Risks

There are several risks associated with any product redesign, including those focused on reducing cost. A few are listed here. The organization should assess if any others exist.

- The organization must assure that all requirements continue to be met after incorporating any cost-reduction-related design changes. Engineering analysis and testing will be required.
- Engineers may resist proposed design changes. If the resistance is based on a requirements compliance issue, it is appropriate. If it is based on a "not invented here" reaction or on turf issues, it is not appropriate. Having an engineer in the cost-reduction brainstorming meetings will help to mitigate

this risk, as will senior management support for what the team is attempting to accomplish.

- The marketing or sales organization should be consulted to assess the effects of proposed changes on product sales, and to assess customer reactions.

- In some cases, the customer may be the design approval authority. This is frequently the case on government programs, and it may also be the case on programs where the organization is building a product to the customer's drawings or specifications. In these situations, approval from the customer is required prior to implementing design changes.

- In some cases, the implementation costs of a proposed design change may outweigh any design-related cost reduction. Careful analysis is required to assess the implementation costs. Design changes should be implemented only after they are deemed financially acceptable.

References

C. Fallon, *Value Analysis*, Washington, D.C., The Miles Value Foundation, 1980.
R.J. Eggert, *Engineering Design*, Upper Saddle River, Prentice Hall, 2004.

22

Requirements Relaxation

The Bottom Line

Requirements should be compared to the existing design for two reasons. In some cases, evolutionary products may contain features not required in the current application. In other cases, the requirements may have been ill conceived. Requirements relaxation or design feature elimination may be appropriate in such cases. In cases where the customer controls the design, customer approval must be obtained to eliminate features no longer required or to relax the requirements. Even if the customer does not own the design, the organization should consider customer reactions prior to taking actions in this area.

Key Questions

Have we assessed all of the features in our current products against their requirements?

Are there any requirements driving our costs that could be reasonably relaxed or eliminated for a cost savings?

Are there any features in our products that the customer doesn't need or care about?

The Requirements Relaxation Road Map

The most important characteristic of any product is that it meets its requirements, but in many cases, the requirements themselves are questionable. In cases where the requirements are unnecessary, too stringent, or no longer valid, it may be possible to eliminate or relax the requirements. If that occurs, the cost will decrease either because rejections related to the requirement will decrease, or a product redesign will eliminate features (and cost) that were necessary to meet the requirement.

Requirements and the Design Process

When designing products, the design should respond to a set of requirements. These requirements can come from the sales department, the engineering department, the customer, or other sources. Requirements are usually based on needs. The needs may be based on aesthetics, performance, reliability, or other factors. In some cases, the needs are determined subjectively. In other cases, the needs are based on quantitative technical analyses. In all cases, assumptions usually drive the definition of requirements, and sometimes the assumptions are wrong. In these cases, the product may have been designed to meet requirements that are too stringent. This presents potential opportunities for requirements relaxation and cost-reducing design changes if the overly-stringent requirements can be identified.

In many cases, new products are evolutions of old products. This means that an older design has been updated to add features, capabilities, or characteristics. Whenever this occurs, every feature on the new design should be evaluated

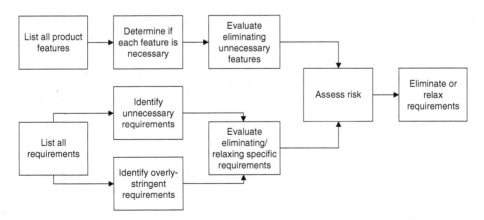

Figure 22.1 Recommended requirements relaxation approach.

in light of the new requirements. Sometimes features inherent to the original design find their way into the evolved design, although the newer product may not need all of the original features. These become candidates for elimination, which will result in a cost reduction.

In aerospace and defense products, design requirements are usually "boiler plated" into the product specifications. This means that in many cases the same design requirements for temperature extremes, shock, vibration, corrosion resistance, etc., are routinely (and sometimes mindlessly) applied to new products.[1] Any overly-stringent or nonrelevant such requirements should be identified and removed during the design phase, but in many cases this does not occur. These similarly become candidates for elimination and cost reduction.

For products already in production, it makes sense to examine all of the original design requirements. If irrelevant or questionable requirements complicate the production process or add unnecessary cost, these may be targets for requirements relaxation. In cases where products are designed to meet government or other client specifications, any proposed requirements relaxation will have to be approved by the client. If the requirements are self-imposed, client approval may not be necessary.

A good way to approach this analysis is to review the drawings and specifications that define the product, and question every requirement and feature. This kind of analysis can reveal unnecessary requirements, requirements that are overly-stringent, and features no longer needed. These situations present opportunities for cost reduction if the requirements can be eliminated or relaxed, or if the features can be eliminated. The questions to ask are:

What is the purpose of this feature or requirement?

Is the feature needed for this application?

If it is still needed, do we need as much?

What are the effects of eliminating or relaxing the requirement?

Requirements Relaxation Examples

A company that designed and manufactured water treatment systems sold one to a client who planned to install it inside a building. The original treatment

---

1. The "$600 hammer" purchased by the Air Force and widely publicized several years ago is a good example. The clamor over this discovery centered on a perception that the supplier was cheating the taxpayers by charging too much. In reality, the cost was real, and it was driven by having to design and test the hammer to inappropriately-specified Air Force requirements.

Figure 22.2 Indoor water treatment system installation. The indoor installation allowed relaxing the temperature and inclement weather requirements, which eliminated many costly design features.

system design was intended for outside installation. It included many features that allowed operation at temperature extremes and in inclement weather. When the manufacturer designed the new system, they were able to reduce the cost by relaxing the temperature and inclement weather protective requirements, which led to the elimination of many unnecessary and costly features.

A defense contractor manufactured tactical lasers for the U.S. Air Force. The lasers were mounted on combat aircraft, and they were used to illuminate targets with laser energy. Smart munitions picked up the reflected laser energy, which guided them to their targets. The lasers also provided range information to the pilot. The original Air Force specification required high laser energy output levels, which were difficult to attain and frequently damaged the laser's internal optical elements. The Air Force found that the laser system could meet its maximum target designation and ranging requirements at an energy level lower than the level originally specified. The Air Force reduced the laser's required energy level, which resulted in a cost savings at the laser manufacturer and in the field due to fewer lasers requiring repairs.

A water company installed a treatment plant to remove nitrates from groundwater. The government required that drinking water have nitrate levels less than 45 mg/L. The treatment plant required salt for regenerating the treatment media, and it generated brine waste that had to be disposed. Both the salt consumption and waste generation volume were proportional to the amount of nitrate the treatment plant removed. The water company's self-imposed requirement was to take the nitrate level to zero. When the water company ran the treatment plant in this manner, it experienced excessive salt consumption and waste disposal costs. The company relaxed its internal requirement such that the treated water had nitrate levels of 40 mg/L, which still met the government requirement. When it did this, salt consumption and brine disposal costs dropped significantly.

Who Should Do This Work

The cost-reduction team can serve as a catalyst here, but when seeking to relax requirements or eliminate features, the bulk of the analysis and testing is an engineering responsibility.

As explained earlier, if the customer has design approval authority, the organization must seek customer approval for any proposed design feature elimination or requirements relaxation. The sales group can help with the interface, but the customer will most likely require inputs from the engineering group to make its decision.

Risks

Risks associated with requirements relaxation are described next.

- If customer specifications or drawings define the product's requirements, customer approval will be necessary prior to relaxing the requirements. This risk can be mitigated by including engineering analyses and test results showing acceptable performance if the requirements are relaxed.

- The reaction to a requirements relaxation may be negative even if customer approval is not required due to a perceived cheapening of the product. The organization should consult the marketing organization to determine if adverse customer perceptions are likely to result from a requirements relaxation.

- The organization should assess all impacts of any proposed requirements relaxation, especially if the requirements relaxation results in an associated design change. Design changes can easily have more effects than anticipated; the organization has to identify all effects to assess their acceptability.

References

J.O. Grady, *System Requirements Analysis*, New York, , Academic Press, 2006.
R.J. Eggert, *Engineering Design*, Upper Saddle River, Prentice Hall, 2004.

23

Tolerance Relaxation

The Bottom Line

Dimensional tolerances specify allowed variability around nominal dimensions. In general, as tolerances become smaller manufacturing costs become greater. The approach used by most organizations for assigning tolerances offers opportunities to reduce cost by increasing tolerances. The organization should identify where tight tolerances increase cost by seeking input from manufacturing, quality assurance, and the suppliers. Increasing the tolerance should not be a default response to a nonconformance, but where it makes sense to do so, tolerances should be relaxed. This should occur only after a rigorous tolerance analysis indicates doing so is acceptable.

Key Questions

How do we assign tolerances?

Do our suppliers or we have any recurring rejections we suspect are induced by needlessly-stringent tolerances?

Are there any areas where our suppliers or we are taking extreme measures to hold tight tolerances?

Do we require drawing changes to relax the tolerance whenever we disposition nonconforming parts "use as is"?

The Tolerance Relaxation Road Map

Before the Industrial Revolution, workers made parts one at a time, and the same worker who made the parts usually assembled them. These early craftsmen cut and filed the parts as necessary to make them fit into the next assembly. There was little or no interchangeability. A part made for one assembly usually would not fit into another assembly.

Eli Whitney realized that a different manufacturing approach would be needed to manufacture enough cotton gins to support America's emerging agricultural economy. Even in the 1700s, Whitney knew that part interchangeability would allow mass production if some workers made parts and others were able to assemble them.

Whitney sold his mass production concept to the U.S. government and he won a contract to manufacture 10,000 rifles. He set up a production operation and built parts for the first 700 only to find that just 14 rifles could be assembled. The parts for the rest would not fit together. Whitney had the right idea, but the manufacturing processes of the era could not produce parts similar enough to guarantee interchangeability. The parts' dimensional variations were too large.

No one knows with certainty who first attained mass-produced-parts interchangeability, but most people agree that it occurred in the early 1800s in the northeastern United States. The first applications were clocks, agricultural equipment, and guns. Originally, the approach consisted of roughing out parts on milling machines and lathes, and then filing the parts to match a template. Engineers soon realized that if they could hold the finished parts' dimensions

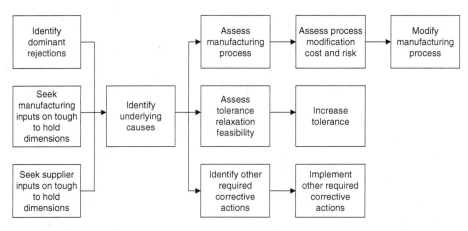

Figure 23.1 Recommended tolerance relaxation approach.

within tight enough limits around the desired dimension, there would be no need to file the parts to fit a template.

Tolerances

This situation resulted in the emergence of drawing tolerances. A tolerance is a plus or minus dimension around the desired dimension. The desired dimension is referred to as the *nominal dimension* and the plus or minus window around the nominal dimension is referred to as the tolerance.

For example, suppose a part has a hole in it with a diameter of 1 inch (the 1-inch diameter in this case is the nominal dimension). The engineer knows it will be impossible to make each hole diameter exactly 1 inch, so the nominal dimension needs a tolerance. It might be ± 0.01 inch, or ± 0.05 inch, or ± 0.10 inch.

It is the engineer's job to specify an appropriate tolerance. The tolerance around this 1-inch hole should be based on fit, function, and the manufacturing method. Here's what this means:

- The part may be part of an assembly with another part fitting into the hole. The tolerances should be specified such that the mating part fits into the hole under all combinations of the hole's tolerance and the mating part's tolerance.

- The hole may not have to interface with another mating part (for example, it might be a vent hole). In a situation like this one, the engineer has great latitude. If the hole is just an air vent, the tolerance can be fairly large. If hydraulic fluid has to flow through the hole, though, the diameter of the hole might be critical to maintaining a desired flow rate or pressure. In this case, the tolerance may need to be tighter. Here, the function of the part governs what the tolerance must be.

- The hole may be drilled, punched, or cast into the part. Each manufacturing process has a capability related to how tight it can control the tolerance. The engineer should compare the desired tolerance to the manufacturing process. If the manufacturing process can hold the tolerance, life is good. If it cannot, either the organization needs to use another manufacturing process that can hold the tolerance or the engineer needs to relax the tolerance. If the engineer has to relax the tolerance, then other parts of the design may require redesign such that the part (and the assembly it goes into) will fit together and work properly.

Tolerance versus Cost

Another important point is that tolerances are generally inversely proportional to manufacturing cost. As tolerances become tighter, manufacturing cost generally increases. From a cost perspective, we want the tolerances to be as loose as possible but consistent with assembly and performance requirements.

Tolerance Determination Approaches

The aforementioned approach describes how engineers are supposed to assign tolerances. Many organizations, however, do not assign tolerances as described here. What usually happens is that an engineer develops the general design concept, a designer (typically an AutoCAD technician[1]) converts the engineer's concept into engineering drawings, and the designer (not the engineer) assigns tolerances. Here are the common approaches most designers use when assigning tolerances:

- In some organizations, the same tolerances are applied to all parts. In other words, everything is assigned a tolerance of, say, ± 0 .05 inch. This makes it easy for the designer to assign tolerances, but because the tolerances are not based on fit, function, or the manufacturing process, many problems in these areas may emerge. This is a poor way to assign tolerances. If your organization uses this approach to assign tolerances, you have many cost-reduction opportunities.

- In some organizations, the tolerance is based on the magnitude of the nominal dimension. For example, dimensions up to 1 inch might get a tolerance of ± 0.001 inch, dimensions up to 5 inches might get a tolerance of ± 0.01 inch, and everything above 5 inches might get a tolerance of ± 0.05 inch. This is another poor way to assign tolerances. If your organization uses this approach to assign tolerances, you have many cost-reduction opportunities.

- In some organizations, tolerances are automatically determined based on how many decimal places are in the nominal dimension assigned by the designer. If the designer specifies a nominal dimension of, say, 1.000 inch or 1.750 inches (three decimal places), the tolerance for both would be ± .001 inch (all three-decimal-place dimensions are assigned a ± .001 inch tolerance). If the designer specifies a nominal dimension of 1.00 or 1.75 inches (two decimal places), the tolerance for both would be ± .01 inch. If the designer specifies a nominal dimension of 1.0 or 1.8 inches (one decimal place), the tolerance for both would be ± .05 inch. While this is slightly better than the preceding two approaches, it is still a poor way to assign tolerances. The tolerances are restricted to fixed steps (± .001 inch, ± .01 inch, or ± .05 inch), and it's not likely these fixed steps correspond to fit or function

---

1. Here, we are using "AutoCAD" in the generic sense. Other software programs (e.g., SolidWorks) are often used in computer-aided drafting applications.

requirements or to process capabilities. It means there may be cases where the parts won't fit together or work, and it means there will probably be cases where the tolerance is tighter than it needs to be for the parts to fit together or work. If your organization uses this approach to assign tolerances, you have many cost-reduction opportunities.

- In some cases, designers believe that assigning a tighter tolerance to a part somehow makes it a higher quality part. In these situations, designers tighten the tolerances in a misguided effort to make it "better." Those designers, intentions notwithstanding, are building a lot of cost into the product. If your organization uses this approach to assign tolerances, you have many cost-reduction opportunities (and the designer needs to be trained on how to assign tolerances).

- In some cases, either the engineers or the designers assess how the parts fit together, what the parts have to do, and how the parts will be manufactured. The engineers or the designers then assign tolerances based on these assessments. This is the way tolerances should be determined.

Engineering drawing tolerances should be assigned based on the last approach (how the nominal dimension and its tolerance will affect fit with the next assembly, functionality, and the manufacturing process's capability to hold the desired tolerance). Modern computer-aided design programs have built-in capabilities for evaluating fit at the next assembly level. The engineers or designers have to consider functionality and the manufacturing process when assigning tolerances.

Tolerance Relaxation Candidates

In most cases, the organization is seeking cost reduction on products in production. There are several ways to identify tolerance relaxation candidates. Here are a few:

- The quality assurance organization can identify specific dimensions that cause frequent rejections. This is similar to the work described earlier in the chapter on reducing scrap and rework. It's always amazing to us how frequently parts are rejected for missing the same dimension. When the same parts show up as dominant nonconformances in a Pareto analysis, the same dimension is usually the dominant cause for the part's rejections. After identifying these items, the next step is to assess the manufacturing process. If the process cannot hold the required tolerance, it is senseless to continue without either relaxing the tolerance (if the design allows it to be) or using a different manufacturing process (one that can hold the required tolerance). Either approach will reduce cost, because the rejections will stop.

- The quality assurance organization can identify all instances in which parts with nonconforming dimensions were dispositioned "use as is." If the parts were acceptable for use with the nonconforming dimension, it will usually be possible to increase the tolerance on the drawings for all future parts.[2]

- The manufacturing organization can identify the specific dimensions they have trouble meeting. In many cases, these tough-to-meet dimensions may not show up as rejections in the nonconformance database because the manufacturing organization is meeting the tolerance, but at a high cost. The same options exist as described earlier: Either the tolerance needs to be relaxed (if the design allows it to be), or a different manufacturing process is required. Again, either approach will reduce cost.

- The purchasing organization can reach out to the supplier community and ask for specific dimensions that are tough to hold. This is a particularly important area to explore, because it's not likely your organization will be able to get this information without asking. If the supplier is rejecting parts, your organization will not see the rejection reports (but you will see increased prices to cover the cost of the supplier's rejections). If the engineering department can relax the tolerance, it should do so. You would be justified in asking for a price concession from the supplier if you do this.

Evaluating Troublesome Tolerances

Once these troublesome tolerances have been identified, the organization now needs to determine how to proceed. The first step should be to determine why the dimensional requirements are not being met. It may be that the machines require maintenance, tools require replacement, the lighting is inadequate, the operator requires training, or the inspector is wrong and the parts are conforming.

If none of these conditions exists and the inspection results are correct, then the conclusion must be that the process is incapable of holding the required tolerance. In that situation, there are three alternatives:

- The manufacturer has to increase the tolerance.
- The manufacturer has to modify the process to hold the required tolerance.
- The manufacturer has to live with the recurring rejections.

---

2. There are exceptions. In some cases, dimensions on assembly-specific mating parts will allow the nonconforming item to be used only on that assembly. In practice, however, dispositioning parts with nonconforming dimensions "use as is" is a tacit admission that the tolerances are too tight.

The third option is really no option at all. The recurring nonconformances are the reason the area is receiving attention.

If the organization wants to relax the tolerance, the designers need to analyze the tolerances on the affected part and all mating parts. This will identify the maximum tolerances that will allow the parts to fit together and meet functionality requirements. Computer-aided design programs make this much easier and far more accurate than when designers had to analyze tolerances manually. If the tolerance can be increased, that is a good way to go, and the cost reductions will occur immediately due to fewer parts being rejected.

If the tolerance cannot be increased, the manufacturer has to modify the process such that the existing tolerances can be met.

Who Should Do This Work

The manufacturing and quality assurance groups should identify areas in which tolerances are causing rejections and extraordinary measures are required to make conforming material. The quality assurance group can do this by examining its nonconformance database. The manufacturing group can probably recite from memory where tight tolerances are causing problems.

The engineering group should assess if questionable tolerances can be enlarged, and by how much.

The purchasing group should solicit supplier inputs regarding unnecessarily-stringent or difficult-to-hold tolerances.

The cost-reduction team should include tolerance relaxation as an area requiring support from the listed groups as part of the overall drive to reduce cost.

Risks

If a tolerance is relaxed without first thoroughly analyzing worst-case tolerance conditions on the part in question and on all interfacing parts, parts manufactured to the increased tolerance may not fit together, or the assembly may not meet performance requirements. The organization can mitigate this risk by assuring the tolerance analysis is thorough.

As mentioned earlier and in preceding chapters, the manufacturer may have to seek customer approval prior to increasing tolerances. If this is the case, the risk of customer disapproval can be mitigated by including a thorough engineering analysis with the tolerance increase request.

It is tempting to ask for a tolerance increase as a default response when rejections occur. The organization has to identify the nonconformance root cause prior to asking for any tolerance increases. It may be that the nonconformances can be eliminated through other corrective actions without the need for a tolerance increase. Even though this may be the case, it still makes sense to explore tolerance relaxations.

References

C.M. Creveling, *Tolerance Design: A Handbook for Developing Optimal Specifications*, Upper Saddle River, Prentice Hall, 1996.

P. Drake, *Dimensioning and Tolerancing Handbook*, New York, McGraw-Hill, 1999.

24

Materials Substitution

The Bottom Line

For a variety of reasons, products are sometimes designed with unnecessarily costly materials. Sometimes this occurs because more modern materials are not yet available, sometimes it occurs because requirements change, and sometimes it occurs due to poor engineering. The organization should brainstorm material substitution opportunities. Where such opportunities are identified, the organization's engineers should evaluate proposed material substitutions to assure compliance with all requirements. The marketing group should evaluate the proposed material substitution from a customer acceptance perspective.

Key Questions

What are the more expensive materials used in our products?

Where can we substitute lower cost materials?

Have new materials become available that might lower our costs?

Are there any places where plastics can be substituted for metals?

The Materials Substitution Road Map

Engineers sometimes specify unnecessarily expensive materials, and replacing these with less expensive materials can significantly reduce cost. In a good design, the lowest cost materials are specified based on environmental, strength, weight, and other requirements. In many cases, though, materials other than those with the lowest cost find their way into products. In some cases, lower cost design materials have become available after the product was designed. Where any of these situations have occurred, cost-reduction opportunities exist if the organization can substitute lower cost materials that still meet the design requirements.

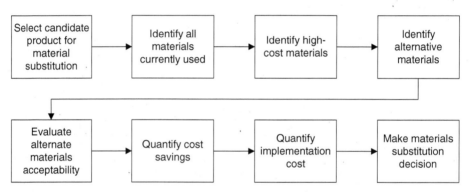

Figure 24.1 Recommended materials substitution approach.

Here's an approach for identifying materials substitution opportunities:

- Identify all materials the product uses.
- Identify which of these are the most expensive materials.
- Evaluate each part made with expensive materials to determine if its requirements can be met with lower cost materials.
- Determine the cost savings associated with changing to less expensive materials.
- Determine the implementation cost of changing to less expensive materials.
- Determine if the savings justify the changes.

Material Substitution Examples

There are many materials and each part's requirements will be unique, so it is impossible to recommend a universal set of material substitutions. Several relatively expensive metals frequently find their way into products. Where these situations occur, we can offer ideas on potentially acceptable alternative materials. Here are a few:

- Titanium is often used where both weight and strength are dominant requirements (e.g., in aircraft applications). If the requirements are such that titanium is the only material that meets the requirements, then it should be used. In some cases, though, engineers may specify titanium just because adjacent parts use the material, or because they are used to working with it. If minimizing weight is not a requirement, any of several steels may be a suitable alternative. If weight is an issue and the strength requirements can be met with other materials, then aluminum or plastic can be considered. In some cases, titanium is specified for aesthetic reasons (examples include jewelry, watches, and high-end bicycles). Even in these cases, though, it may be that another material can be used if it is acceptable in the marketplace.

- Stainless steels are often specified for corrosion resistance or aesthetic reasons. Where corrosion resistance is a requirement, metallurgists can identify the lowest cost stainless steel that meets the corrosion resistance and aesthetic requirements. Alternatively, a lower cost painted carbon steel or aluminum might meet the corrosion resistance requirements. If a part is made of stainless steel and corrosion resistance is not a concern, it may be that any of several lower-cost carbon steels or aluminum with an appropriate surface treatment could be used instead. Modern plastics may allow meeting the part's strength requirements; if this is the case, a lower cost plastic could be substituted.

- Aluminum materials are often specified where weight is a concern. In some industries, engineers get used to working with aluminum, and in these situations aluminum may be specified where a lower cost material could be used instead. If weight is not a concern, steel or plastic could possibly be used. Additionally, there are many different kinds of aluminum and the costs for these vary widely. If aluminum is required, consulting a metallurgist may result in the identification and selection of a lower cost aluminum.

- Steel pipe and copper pipe are frequently specified for water, sewage, and other liquid transport systems. For pipe applications where high pressure is not a requirement, it frequently is far less expensive to use plastic tubing or PVC pipe instead of steel or copper pipe (this is true for pipe diameters of up to about 12 inches; beyond that point, the costs are about the same). If PVC pipe is used instead of steel or copper, the savings extend beyond just the pipe. Other associated parts (elbows, tees, and other fittings) can be plastic. The installation labor for plastic pipe is significantly lower than it is for steel or copper. Plastic pipe or tubing usually does not need to be painted, as does steel pipe.

- Steel tanks and pressure vessels are frequently specified where large fluid containers are required. In some cases, fiberglass, plastic, or other materials can be used instead of steel. For low-pressure applications, molded plastic tanks are far less expensive than steel tanks. For higher-pressure applications, it may be possible to use filament-wound fiberglass vessels instead of steel pressure vessels. In addition to costing less, in many cases filament wound vessels are actually stronger and weigh less than steel vessels. Their lower weight will reduce shipping and installation costs as well.

- Structural lumber is frequently specified for construction, housing, and other applications. In many cases, engineered lumber products (i.e., composite lumbers or particle lumber) will meet the requirements at a much lower cost.

This is not intended to be an exhaustive list. It is intended to stimulate thinking. Material substitution makes an ideal topic for a brainstorming session.

Who Should Do This Work

The cost-reduction team can, as recommended earlier, take on the topic of material substitution in a brainstorming session. After potential material substitutions have been identified, the engineering department should evaluate the acceptability of each.

Risks

If the customer owns the design, any proposed material substitutions will require customer approval. A complete engineering analysis showing the new material's acceptability under all load and environmental conditions can mitigate this risk.

If your organization owns the design, you should still have the engineers assess the new material's acceptability under all environmental conditions.

Sometimes material substitutions can create the impression of lower quality. This might occur, for example, if plastic is substituted for steel. Even though the product may be as good or better than the original design, customer perceptions may make a material substitution unacceptable. For this reason, it is always good to get the sales group's input on any proposed material substitutions.

References

L.D. Miles, *Techniques of Value Analysis and Engineering*, New York, McGraw-Hill, 1972.

S. Kalpakjian, *Manufacturing Engineering and Technology*, Boston, Addison Wesley Publishing Company, 1992.

C. Fallon, *Value Analysis*, New York, McGraw-Hill, 1980.

25

Packaging

The Bottom Line

Packaging can be a significant part of the product's cost, and packaging redesigns can result in significant cost reduction. Packaging designs should be assessed in the same manner as product designs with an objective comparison of the packaging design to its requirements. Brainstorming is a good way to identify lower cost packaging options, as is retaining outside expert assistance.

Warranty data will indicate if the packaging is under-protecting the product; if this is the case, packaging expense may increase but overall costs should decrease due to fewer warranty returns. In many cases, packaging can be reused for additional savings.

Key Questions

Who designs our packaging?

What is our packaging design process?

Have we considered lower cost packaging designs?

Do we have a packaging expert either on staff or available to us?

211

Do we have any opportunities to reuse packaging either in-house or with our suppliers
   or customers?

What do we do with discarded packaging materials?

The Packaging Cost-Reduction Road Map

A product's package can be a significant part of its cost. Overdesigned packages add unnecessary cost. Underdesigned packages add warranty cost. This
chapter presents an approach for determining if packaging is overdesigned or
underdesigned. It also suggests approaches for reducing packaging costs.

Packaging design is heavily influenced by marketing, functional, environmental, and cost considerations. Packaging can have a huge impact on how
well a product sells, and for this reason, the shape and print on a package is
often used to sell the product. The package's functional requirements drive its
design (the package protects the product, it may have requirements that allow
the product to be stacked, it may allow the purchaser to pour from the package,
it may enhance the product's shelf life with built-in desiccants or flavor enhanc-
ers, and it may make the product tamperproof). The packaging may have been
designed to meet regulatory or self-imposed environmental requirements (the
package may be constructed of recycled materials, it may have to be capable of
being recycled, it may be biodegradable, etc.). Finally, cost is a consideration in
packaging design. Our objective is to minimize packaging costs while allowing
the package to meet all of its other requirements.

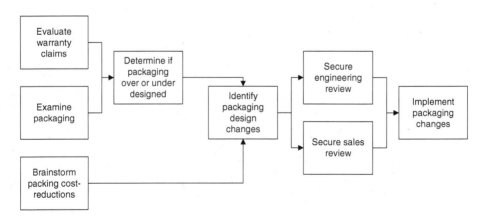

Figure 25.1 Recommended packaging cost-reduction approach.

Packaging Adequacy

In seeking to minimize packaging costs, consider the problem from two perspectives:

- If the packaging is overdesigned, it is providing too much protection to the product. It may be possible to reduce cost by reducing some of this protection.
- If the packaging is underdesigned, it is not providing enough protection to the product. In this case, it will be necessary to redesign the product to provide more protection, which may increase the packaging cost. If the packaging cost increase results in less product damage, though, the overall cost will decrease.

Examining the packaging and reviewing the organization's quality and warranty data are good inputs for assessing the issue of overprotection versus underprotection.

Overdesigned Packaging Indicators

A simple packaging examination with the engineers can be revealing. For example, consider how new personal computers are packaged. In many cases, each component is wrapped in a plastic bag that is sealed with a wire wrap. The plastic-bag-protected components fit into Styrofoam moldings. These are then enclosed in cardboard boxes, which are in turn loaded into larger cardboard boxes. Within the large cardboard box, additional cardboard and Styrofoam separators segregate and restrict component movement. The cardboard box containing the entire system is made of heavy corrugated cardboard, and the exterior is completely covered with high-gloss printing advertising the computer's features.

Is all of this packaging necessary? Do all of the components need to be in plastic bags? Do the component Styrofoam castings need to be enclosed in their own cardboard boxes? Is the weight of the cardboard (its ply) necessary, or will a lighter grade suffice? Do we need to cover the exterior cardboard box with high gloss printing, especially since the computers are stored in warehouses until the consumer makes the purchase? What are the alternatives to the current packaging approach?

Sometimes a simple examination and a bit of common sense can identify packaging cost-reduction opportunities. A product's packaging is another good topic for a brainstorming session.

Warranty Data Indicators

Reviewing the organization's quality assurance and warranty data can also reveal opportunities. For example, if the organization experiences frequent product damage and the damage is attributed to the packaging, the packaging is not doing its job. Although a packaging redesign may increase cost, resultant reductions in material handling and transportation damage may result in an overall lower cost (and increased customer satisfaction). Another option is to redesign the product being protected by the package, such that the product is more robust and requires less protection.

On the other hand, if there are no warranty claims or other nonconformances related to the packaging, it may indicate either that the packaging is adequate or that it is overdesigned. By itself, a lack of packaging-related damage suggests (but does not prove) that the packaging could be modified to offer less protection. Further analysis is required.

Packaging Cost-Reduction Ideas

The previous discussion has focused on consumer goods at the retail level. Many products are packaged by suppliers for delivery to manufacturers. These areas also offer opportunities for packaging cost reduction.

Here are several suggestions for packaging cost reductions:

- Small products provided in separate packages may be candidates for bundled packaging. This is especially true in a business-to-business environment, but it may also apply in a retail environment (it depends on the product). We observed a situation in which a metal lanyard supplier packaged each lanyard in a cardboard box, and then put several hundred packaged lanyards in a larger cardboard box. The manufacturer to whom the lanyards were provided employed a person whose job consisting of dumping all of the packaged lanyards onto a workbench, removing the lanyards from their individual boxes, and then putting the lanyards back into the larger box. He then moved the box of loose lanyards to the assembly line, where technicians removed the lanyards as they needed them. The lanyards could not tangle (they were not long enough), and they were robust enough that they could not damage each other during transport. The obvious cost reduction was to eliminate packaging the lanyards separately at the supplier. This eliminated the cost of the individual lanyard boxes, the printing on these boxes, the labor to put the lanyards into the individual boxes, and the labor to remove the lanyards from the individual boxes. After implementing this packaging change, the lanyard manufacturer could place more lanyards in the larger box, which reduced shipping costs.

- Packaging material selection will drive cost. We discussed material substitution as a product design cost-reduction consideration in the previous chapter; the same analysis is applicable to packaging design. The organization should assess plastic, cardboard, expanded foam, wood, steel, and other materials to identify which packaging material best meets all requirements at the lowest cost. For liquid products, the trade study should evaluate glass, plastic, sealed cardboard, metal, and perhaps other containers.

- For larger products, packaging may be unnecessary. Large material coils, for example, do not usually need to be packaged. They can be labeled and shipped directly.

- Sometimes larger products can be palletized and shipped on the pallet instead of being boxed.

- Many products are palletized and shrink wrapped. Sometimes the shrink-wrapping is unnecessary. If it is necessary, sometimes a thinner shrink-wrap material or fewer wraps will work.

- In many cases, packaging can be reused. A few decades ago, it was common for soft drink and milk bottles to be returned for reuse (in fact, soft drink suppliers paid for each returned bottle). The model is a good one to consider, particularly in a business-to-business sales environment. There are many trades to be made in considering the best approach if the packages are going to be reused. Cardboard boxes, for example, may have the lowest initial packaging cost but they may only survive three uses. Plastic or metal containers may have higher initial cost, but their service life (the number of times they can be reused) lower the overall life cycle cost. When the U.S. government procures munitions or other expensive items, the packaging cost for these items can be quite high. The packages are reused many times, though, and as a result, the overall cost is lower.

- In many cases, packaging materials can be recycled. This is potentially true for all packaging materials, but it is especially true for cardboard containers. Many smart manufacturers compress the cardboard boxes their suppliers use for packaging and sell the cardboard materials to recyclers. One company used to pay a hauler to take its discarded boxes away; when the company discovered the hauler was selling the used cardboard to a recycler, it contacted the cardboard recycler directly. The cardboard recycler offered to pay the manufacturer for the cardboard and pick it up for free. The manufacturer eliminated its discarded cardboard hauling costs and realized a revenue stream from selling the recycled cardboard.

- Film overwrapping is often a good approach for containerizing multiple items. The concept here is that several items are stacked and then wrapped in printed paper with diamond-point folds held in place with an adhesive (think of gift wrapping). The concept can be seen on multiple packages of bar soap. If a clear film is used, it takes advantage of the film-overwrapped items' printing (eliminating the need to pay for printing on the film overwrap). This approach is a good one where the package contents are rectilinear in shape and of relatively light weight.

- Returning to the computer example mentioned earlier, evaluating the need for printing on the package's exterior can frequently reveal opportunities. Fancy printing may be necessary for marketing reasons, but in many cases, it is not. This is especially true for larger items that are stored in warehouses until the customer purchases the item (e.g., the computer discussed earlier in this chapter), or in a business-to-business sales environment.

- The use of pre-printed labels versus printing directly on the packaging materials may offer cost-reduction opportunities. The analysis should consider the cost of the label versus the cost of printing on the packaging materials.

- The packaging's ink and color selection affect printing costs. Evaluating the use of alternative inks and using fewer colors can reduce cost.

- Cardboard boxes are widely used for packaging products. In some cases, a lower cardboard ply can be incorporated, lowering the packaging cost. This will also lower the overall package weight and its outside dimensions, and these factors may be significant enough to lower the shipping cost.

Reusable Packaging

Packaging materials can often be reused for material handling within a manufacturing facility. Obviously, packaging made of durable and semi-durable materials (metals or plastics) are candidates for reuse, although it is surprising how many organizations do not reuse these items. Even nondurable items (e.g., cardboard containers or Styrofoam) can often survive several usage cycles, thereby reducing the cost of replacing these items after each use. A quick way to gage if you have candidates for reuse is to simply look in the plant's dumpsters. If there are packaging items in there, there are opportunities for cost reduction.

Who Should Do This Work

The opportunities for reducing packaging costs are significant. This is an area where brainstorming by the cost-reduction team and others helps. Consultants can also help. There are only a few schools offering degrees in packaging engineering, so there are not a lot of engineers with this special skill set. It may make sense to seek expert help in this area.

Risks

Packaging can have an enormous impact on sales, so it is critical that the marketing group have an input into any decisions to modify a product's package.

Packaging requirements are sometimes dictated by customers, so if changes are proposed, customer approval may be required. Engineering analyses and tests supporting a proposed change can mitigate the risk of a customer rejection.

Reusable packaging can spread contaminants, as mentioned earlier in the chapter on material-handling damage. If you plan to reuse packaging, you should assess the likelihood of contaminant migration via the package and take appropriate steps to prevent this from occurring.

References

S. DuPuis, and J. Silva, *Package Design Workbook: The Art and Science of Successful Packaging*, Beverly, Rockport Publishers, 2008.

J. Foster, *For Sale: 200 Innovative Packaging Designs*, New York, HOW Books, 2008.

Part V

Overhead

The overhead functions are another area with cost-reduction opportunities. In the context of this book, the term *overhead* refers to overhead, general, and administrative expenses. Overhead cost reduction, the fifth and last category addressed in this book, has chapters focused on reducing cost in the following areas:

- General overhead expenses.
- Travel.
- Inspection.

26

General Overhead Expenses

The Bottom Line

Overhead costs are significant, and reducing these costs must be a key part of any cost-reduction effort. The organization should establish an activity-based overhead budget and track performance to it. The headcount in overhead groups should be established in the same manner as that in direct charge groups. This approach consists of identifying tasks, determining the time required for the tasks, and converting the summed time to a headcount. Specific overhead cost-reduction targets include management span of control, headcount, utilities, facilities, gasoline, cell phones, and using competition for overhead equipment, supplies, and service.

Key Questions

Do we have an overhead budget, and do we monitor performance to it?

How do we determine headcount in the purchasing, quality assurance, and other overhead departments?

How many organizational layers are there from a worker in the shop to the president?

Who gets a company cell phone?

Do we need a facility as big as the one we have now?

Have we recently assessed the necessity of each overhead expense?

The Overhead Cost-Reduction Road Map

General management, sales, procurement, quality assurance, and other areas are frequently designated as overhead or general and administrative functions. In addition to these areas, many expenses (such as office supplies, utilities, etc.) are in the overhead category. Generally, the overhead and general administrative categories include any expense that is not directly and uniquely tied to a product line. As mentioned earlier, in this book we are including all of these expenses in the overhead category.

Overhead Budgets

Let's talk about establishing an overhead budget first. If your organization doesn't have a budget, you should prepare one. When establishing or modifying an existing overhead budget, the basic questions are:

- What are our overhead activities?
- What should these overhead activities cost?
- Should we perform the overhead activities in-house or outsource them?
- When will the expenses occur?

Activities and expenses that are normally categorized as overhead include the following:

- Facilities
- Utilities
- Postage
- Advertising
- Entertainment
- Travel
- Office supplies
- Coffee
- Fuel
- Tools
- Shop supplies
- Sales
- Calibration
- Quality Assurance
- Human Resources

- Finance
- Purchasing
- Management.

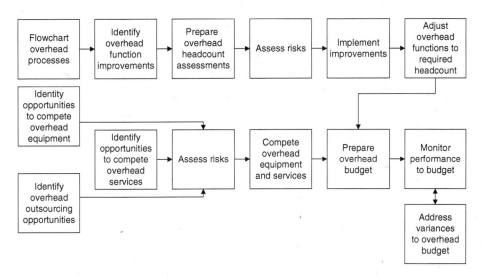

Figure 26.1 Recommended overhead cost-reduction approach.

When assessing the costs in each of these areas, the intellectually lazy approach is to examine historical data and project it forward. Sometimes companies take a challenge when they do this (e.g., by reducing the past year's budget by, say, 3 percent), but this is also a lazy and expensive way to budget. Basing the budget on prior history can build in unnecessary cost. A far better approach is to use a technique called *activity-based* or *zero-based budgeting*. Using this approach, the organization identifies the tasks and then estimates the costs of each. This is an activity-based cost estimate, and it is the preferred approach for developing a more accurate (and generally much lower) budget.

A great way to do this is with an Excel spreadsheet, listing each cost category in the right column of the matrix and the month across the top row. Table 26.1 shows a sample overhead budget prepared in this format.

Immediately after the end of the month, the finance department should provide reports to each department manager showing the actual overhead expenditures on both a monthly and a year-to-date cumulative basis.

With these data, it is a simple matter to prepare a chart showing the cumulative budget line versus the cumulative actual expenditures line (see Figure 26.2). If the cumulative actual expenditures line exceeds the cumulative budget line,

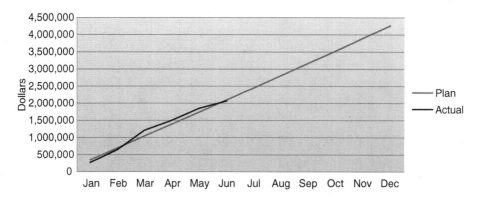

Figure, 26.2 Overhead budget performance. This format shows planned cumulative overhead expenditures as well as actual cumulative overhead expenditures. It shows the organization how it is performing against the budget with a single chart.

the organization is overrunning its budget. This approach allows the organization to monitor status against the overhead budget, and aggressively attack variances to the budget as they occur.

Some organizations require their managers to predict the next three months' expenses every month. This provides a look ahead to alert the organization to looming budget overruns. It also forces each manager to recognize what the budget is.

Table 26.1 and Figure 26.2 show the overhead budget at a company level. It makes sense to have similar spreadsheets and charts at the department level to allow each manager to monitor their performance to the budget.

We mentioned earlier that the finance group should provide actual cost information to the department managers shortly after the end of each month. Good managers won't need this; they will already know what expenses their departments have incurred. The budget information from the finance department should be a backup. Relying exclusively on cost information from the finance group is like driving using the rearview mirror. If you overrun the budget, by the time you get the information from the finance people it is already too late. If department managers track expenses as they are incurred (and they should), they will know if they have a looming budget issue and they can take action to avert it. The three-month look ahead also helps avert budget overruns.

Reducing Overhead Costs

In addition to preparing and using an overhead budget, the organization can assess and attack its current overhead costs simply by preparing a Pareto

Table 26.1 Overhead Budget. This Budget Shows Planned Overhead Expenses for the Year

Cost Element	Overhead Budget											
	Jan	Feb	Mar	Apr	May	Jun	Jul	Aug	Sep	Oct	Nov	Dec
Facilities	15,750	15,750	15,750	15,750	15,750	15,750	15,750	15,750	15,750	15,750	15,750	15,750
Utilities	6,500	6,500	6,500	3,400	3,400	7,000	7,000	7,000	3,400	4,000	6,500	6,500
Postage	950	950	950	950	950	950	950	950	950	950	950	600
Advertising	15,750	15,750	15,750	15,750	15,750	15,750	15,750	15,750	15,750	15,750	15,750	15,750
Entertainment	1,500	1,500	4,500	1,500	1,500	4,500	1,500	1,500	4,500	1,500	5,500	4,500
Travel	31,200	31,200	31,200	31,200	31,200	31,200	31,200	31,200	31,200	31,200	45,000	26,000
Office supplies	4,800	4,800	4,800	4,800	4,800	4,800	4,800	4,800	4,800	4,800	4,800	4,800
Coffee	750	750	750	750	750	750	750	750	750	750	750	750
Fuel	2,500	2,500	2,500	2,500	2,500	2,500	2,500	2,500	2,500	2,500	2,500	2,500
Tools	7,800	7,800	7,800	7,800	7,800	7,800	7,800	7,800	7,800	7,800	7,800	7,800
Shop supplies	33,250	33,250	33,250	33,250	33,250	33,250	33,250	33,250	33,250	33,250	33,250	33,250
Sales	73,724	73,724	73,724	73,724	78,147	78,147	78,147	78,147	78,147	78,147	78,147	78,147
Calibration	6,678	6,878	7,085	7,297	7,516	7,742	7,974	8,213	8,459	8,713	8,975	9,244
Quality assurance	50,208	50,208	50,208	50,208	50,208	50,208	50,208	52,467	52,467	52,467	52,467	52,467
Human resources	12,660	12,660	12,660	12,660	12,660	12,660	12,660	12,660	12,660	12,660	12,660	12,660
Finance	28,580	28,580	28,580	28,580	29,723	29,723	29,723	29,723	29,723	29,723	29,723	29,723
Purchasing	32,000	32,000	32,000	32,000	32,000	32,000	32,960	32,960	32,960	32,960	32,960	32,960
Management	22,540	22,540	22,540	22,540	22,540	22,540	22,540	22,540	22,540	22,540	22,540	22,540
Total	347,140	347,340	350,547	344,659	350,445	357,270	355,463	357,961	357,607	355,461	376,023	355,942

analysis of these costs and seeking cost reductions in the dominant categories. The organization can also seek reductions in areas commonly known to be overhead cost drivers. These are listed here.

- Headcount, when left unchecked, has a tendency to grow. One approach for assessing how many people are needed in each overhead category is to identify all of the necessary tasks, assess how much time these tasks should take, and convert this to a required headcount. This is similar to the approach outlined earlier for direct labor headcount. Sometimes it helps to ask each person what he or she does; at other times, it may make sense to list the tasks people perform without asking them about it.[1] Whichever approach is used, the person preparing the budget can then categorize the tasks into required tasks, nice-to-have tasks, and tasks that are not required. In many cases, a portion of the tasks is either not required or is nice to have. Eliminating some or all of this work reduces the headcount requirement.

- Another important part of reducing overhead headcount is to examine how required tasks are performed. Flowcharting the process, as outlined earlier, may identify opportunities for simplification. Establishing time standards may also make sense. The cost-reduction techniques discussed throughout this book are not limited to the manufacturing work centers; many are equally useful in assessing overhead work content and efficiency.

- Management span of control and staff support often present cost-reduction opportunities. Like headcount, these things tend to move in the wrong direction if not closely managed. Some studies suggest that the optimum number of people reporting to each manager should be seven; other studies have suggested it should be as high as twenty. We believe generalizations like this are meaningless, and that the optimal number of people reporting to each manager depends on the work and the people doing it. Our belief is that if there are more than four or five layers from a shop floor worker to the general manager, something is wrong and the company has too many bosses. If this is the case in your organization, there are probably opportunities to reduce overhead costs by flattening the chain of command.

- Look for the assistants. Sometimes it seems like everyone needs an assistant, when in reality few people do. In many cases, these positions are candidates for elimination.

- Sales organizations are almost always assigned to an overhead account. Tracking performance by salesperson is a good way to keep the sales staff

---

1. Asking employees what they do is difficult to do without raising concerns about job security. Also, some employees may not be able to articulate what they do.

motivated, and it's also a good way to identify salespeople who are not productive. If a salesperson is not selling or is selling very little, it's hard to justify keeping the position. Tracking performance by salesperson is a good way to find cost-reduction opportunities.

- Overhead equipment costs can be reduced by making the equipment suppliers compete. Sometimes renting is less expensive than buying; sometimes buying is less expensive. The organization can request bids from competing suppliers with both sale and lease options for copiers, printers, computers, and other overhead equipment items.

- Office supplies, postage, coffee, and other overhead costs can similarly be competitively bid. All of the techniques discussed in the chapter on supplier competition are applicable here.

- Utilities costs are significant. Installing thermostats with limited access, making sure the lights, heaters, and air conditioners are turned off when people go home, making sure phones are only used for company business, and other commonsense measures will reduce these expenses. More focused steps include using high-efficiency light bulbs, identifying high-energy-consumption equipment (and taking steps to lower energy consumption), and voluntary electrical shutdowns. Many electrical power companies have programs in which customers agree to let the power company shut down power during unusually high demand periods in exchange for significantly lower energy rates. It's rare that the power company actually exercises this authority, so the trade-off for most companies is a good one.

- Gasoline expense is significant for many companies, particularly in light of the run-up in fuel prices over the last few years. For larger organizations, it may be possible to negotiate lower fuel prices for longer-term commitments. In all organizations, maintaining mileage logs to assure company vehicles are being used appropriately will hold the line on excessive fuel consumption. One organization actually plotted miles per gallon for each of its drivers and offered a reward to the driver attaining the best fuel mileage. This reduced that organization's fuel costs significantly. Limiting trips, combining trips, and switching to vehicles with better fuel efficiency also help.

- Cell phones can be a significant overhead expense. Recommendations for cutting cell phone expenses include focusing on who gets a phone, using the phones for business purposes only, negotiating better rates, and evaluating usage. In many cases, organizations have no criteria for determining who gets a cell phone, and phones are issued to almost everyone. Company cell phones should go to employees who need them for business purposes. The organization should evaluate usage when it receives the cell phone bill. People who rack up significant time on personal calls should be advised not to do so. Some companies have experimented with employees using personal phones for company business and expensing the phone calls, but this usually doesn't work well because of the administrative costs of processing the expense reports. As is the case with other services, your company can reduce costs by competing the service and negotiating lower rates.

- Facilities (the property and the buildings) are overhead items, and their expense is usually significant. Evaluating leasing versus buying when selecting a facility will reveal which is less expensive. The organization should also evaluate how much space it needs with a fresh perspective (in other words, defining how much space is needed based on the process, not on the current facility's size). If the factory and office space requirements are smaller than the current facility, there are a couple of options. One is to evaluate relocating to a smaller facility[2]; the other is to sublet a portion of the factory to another business. Either approach can significantly reduce facilities-related overhead cost.

- Outsourcing should be considered for many overhead functions. Many organizations outsource their computer and information technology work, finding it is less costly than employing computer technicians. Many organizations outsource their payroll functions for the same reason. In the quality assurance area, outside calibration labs are frequently more economical than employing an equipment calibration technician and maintaining a calibration lab. Many manufacturers outsource their sales function, relying on manufacturer's representatives. When evaluating shifting an overhead function to an outside supplier, the analysis is similar to the make-versus-buy analysis for items used in manufactured products.

- In some cases, overhead costs can be shifted into the direct cost category. The organization will still incur the cost, but direct charge costs are often more closely managed than are overhead costs so the cost is likely to decrease if it is moved to a direct charge category. Some organizations include the quality assurance functions and purchasing in the direct charge category.

Who Should Do This Work

This is an area in which the department managers are best suited for the budgeting process, and to assess and make the actual cost reductions. The cost-reduction team should develop cost-reduction recommendations and support the overhead cost-reduction effort.

---

2. This evaluation needs to include the costs of the move, including disruptions to the business, transport costs, loss of employees due to potentially longer commutes, etc. The upside is that in addition to a smaller facility, the facility can be located in lower cost areas. Rental costs per square foot can vary by as much as 50 percent over a thirty-mile distance in some areas.

Risks

Reducing overhead expenses has associated risks:

- When assessing headcount, employee morale may be affected. That's not a reason to avoid doing so, but the organization needs to manage the risk and minimize the impact on the remaining workforce. Whenever the organization considers headcount reductions (for either direct or overhead staff), the assessment should proceed quickly. To the extent possible, if there is to be a layoff it should be done once. Few things are more demoralizing to a workforce than a series of ongoing layoffs.

- Attempts to understand overhead job content may be met with resentment. This is similarly not a reason for avoiding the task. In some cases, making your own assessment or having the group manager make the assessment without talking to the employee might be a better idea. If the group manager resists requests to do so, then the organization has a different kind of problem. Higher-level management support may be required.

- When identifying overhead job content, identifying all tasks is critical. Failure to do so when estimating the required headcount could result in more work than can be handled by the remaining staff. This risk can be mitigated by diligence in identifying all tasks.

- Reducing costs in some of the areas identified in this chapter may adversely affect morale. In one organization, for example, there was a near-revolution when the company attempted to reduce what it spent on coffee. Sometimes this risk can be managed by meeting with employees and explaining the reasons for the cost reduction. Sometimes if the expense is small and the employee reaction is severe, it's better to bite the bullet and incur the expense. It's a management call.

- Tracking performance by salesperson will almost certainly evoke a reaction. Sales people often have big egos (it's something needed to be successful in sales), so assessing their performance may not go over well. As is the case with the previously-mentioned tasks, this is not a reason to avoid doing so. Sales people are usually handsomely compensated, so if a salesperson is not performing, the need to make a cut is even more pronounced.

- Attempts to manage utility costs need to consider the operation. An organization that implemented a timer to cut electricity at night found that ovens used for curing products were supposed to stay on, but the timer cut power to everything. This company had to scrap a day's production because it failed to consider this when implementing the change.

- When making decisions to outsource work currently done in-house, the transition needs to be managed. Continuing the work in-house while the transition occurs can mitigate the risk of a discontinuity.

- When switching overhead service suppliers, the transition needs to be managed. Continuing to use both suppliers for a brief period can mitigate this risk, but this may not always be possible. Careful planning will also help to mitigate this risk.

References

J. Olson, *The Agile Manager's Guide to Cutting Costs*, Beaverton, Velocity Business Publishing, 1997.

J.A. Brimson, and J. Antos, *Driving Value Using Activity-Based Budgeting*, Hoboken, Wiley, 1998.

27

Travel

The Bottom Line

Business travel is expensive. Reducing travel costs requires planning trips well in advance when possible, taking only those trips that are necessary, and controlling expenses when traveling. The organization can reduce costs for many required travel services through negotiation and group rates. The organization should prepare a travel budget in the same manner as it prepares other budgets. Superfluous travel should be eliminated, but the organization should not be afraid to travel when it is necessary to do so.

Key Questions

Do we have travel guidelines?

Do we have a travel budget?

Who approves travel plans?

How do we make sure each trip is necessary?

What are we doing to minimize airfare and other travel expenses?

What trade shows do we attend, and what has resulted from our attendance?

The Travel Cost-Reduction Road Map

Travel by the sales staff, buyers, and others on overhead is such a significant cost that it requires separate consideration. In many cases, business travel is required. In other cases, other less-expensive approaches (teleconferences, e-mail, etc.) will work just as well. Part of the problem is that some people just like to travel (especially when someone is paying for it), and they may allow this sentiment to drive up the organization's costs.

Business travel is expensive. Airlines and hotels charge substantially more for trips made only days in advance (which is usually how business trips are scheduled). Rental car companies charge exorbitant rates for their cars, insurance, and fuel. Airport parking fees are high. These costs (along with meals and entertainment expenses) can easily make even a short trip cost thousands of dollars, so it's important to take steps to keep these costs down.

Several recommended approaches for reducing travel costs follow.

- The first step in reducing costs associated with travel is to assess if a planned trip is necessary. Defining a trip's purpose prior to traveling often makes it obvious that the objectives can be met in other less expensive ways. In many cases, there is a value in meeting face-to-face with customers or suppliers, but if that is not a factor and the trip's objectives can be met without traveling, scrubbing the trip will result in significant savings.

- Given that the trip is necessary, it should be scheduled as far in advance as possible. One reason for doing this is that the situation may change, making the trip unnecessary. Another reason is that air travel is usually much less

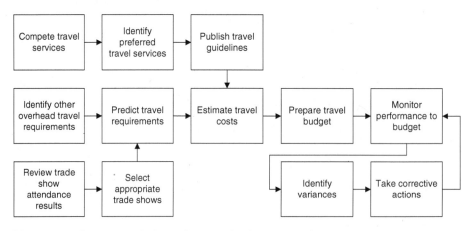

Figure 27.1 Recommended travel cost-reduction approach.

expensive if scheduled well in advance. The airlines love people who schedule their trips a day or two before they travel.

- Many people automatically choose to fly when they have to travel. If the distances are not too great, it might make more sense to take public transportation or drive. Air travel is not the great time saver it once was. The time required to get to the airport, park, go through security, get on the airplane, get off the airplane, claim bags, go to the rental car agency, leave the airport, and drive to the final destination adds hours to any flight, and that's if the airplane leaves on time (and they often do not). In many cases, driving will actually take less time. The tipping point from a time perspective seems to be about 250 miles. If the distance is less, it probably makes more sense to drive and the cost will probably be lower.

- Some organizations fly their more senior personnel first class. This is expensive and wasteful. First-class tickets can cost as much as ten times the amount a coach ticket costs. First-class air travel is quite a bit nicer than flying coach, but it is a luxury and not a necessity. If you are seeking ways to reduce travel costs, this is a good one. Eliminating this wastefulness at the executive level sends an appropriate message to everyone else in the organization as well. The message is that the organization is serious about cost reduction.

- A few years ago, many companies had in-house employees who did nothing except arrange travel for other employees. Many organizations then moved toward using outside travel agencies. Today, most companies have their employees arrange their own travel because it costs less. Some might argue that the employees' time is better spent doing their regular jobs, but the time required to arrange a trip is minimal (especially now that it can so easily be done on the Internet).

- Many organizations negotiate rates with hotels, taxi services, and rental car companies. Hotels often offer substantially reduced rates to gain exclusivity with a client. Sometimes they'll include free dining if they have a restaurant.

- Rental car companies may offer lower daily rates (and refueling at more sane gasoline prices) in exchange for exclusivity. Limo and taxi services will similarly offer lower rates for corporate clients. It's there for the asking; if your organization is not asking, it is missing a great cost-reduction opportunity.

- Airport parking is another exorbitant expense that can be significantly lowered by using remote parking or non-airport-affiliated commercial parking lots. At one local airport, the remote parking rate was one-fourth what it cost to park at the terminal. Simply directing employees to use the remote parking facilities can reduce this expense significantly.

- When employees travel, it is reasonable to expect them to travel on the same day as the visit if the trip is not a long one. In some organizations, the culture is such that employees take a day to travel (always on a weekday), they conduct business the next day, and then they return home the next day (again, always on a weekday). This is abusive.

- It makes sense to require receipts for incidental expenses, and to establish a daily per-diem amount for travel. This motivates employees to seek reasonable accommodations and dining. Simply establishing a per-diem rate without requesting receipts often results in costs exactly equaling the per-diem amount. Establishing a per-diem rate and requiring receipts often results in cost reductions.

- Attending trade shows costs thousands of dollars. They often result in nothing more than added cost. Trade shows should be viewed from the perspective of what they do to help develop business, and not as an opportunity to travel and party on the company's dime. If trade shows are not bringing in new business, their cost is hard to justify. Reducing or eliminating trade-show participation (if your attendance does not result in new business) will reduce costs significantly.

- As is the case for other expense categories, it is a good idea to establish a budget for overheard travel. The budget should not be based on what the organization spent last year; it should be based on the trips it plans to take this year. Each trip and its associated cost should be identified as part of the budgeting process. When trips have to be justified in advance and their costs included in a budget (as opposed to simply traveling and incurring the cost), some of the travel seems pretty silly.

- The organization should prepare and publish travel guidelines. The guidelines should specify preferred service providers, per-diem amounts, and other requirements to communicate company travel expectations.

Who Should Do This Work

The organization's senior management needs to define the company's travel policy, especially as it pertains to such things as first-class air travel, per-diem allowances, and the other cost-reduction opportunities described in this chapter. The finance group, the purchasing group, and the human resources group should all provide inputs. The purchasing group should take the lead in negotiating lower rates for air travel, rental cars, and hotels.

Risks

There's not much downside in attempting to reduce travel costs. The organization should exercise diligence in assessing travel needs and it should minimize costs as outlined in this chapter, but people in the organization should not be afraid to travel when it is necessary to do so. It would be silly to miss business development opportunities in a quest to reduce travel costs. This risk can be managed by exercising good judgment.

References

R. Montgomery, "Reducing Travel Costs For Your Company," *San Fernando Valley Business Journal*, September 25, 2006.

R. Collis, *Survivor's Guide to Business Travel*, London, Kogan Page, 2002.

28

Inspection

The Bottom Line

Inspection costs are significant, yet inspection adds nothing to the product's value. Approaches for reducing inspection costs include improving where inspections occur, eliminating redundant inspections, and automating inspections to lessen susceptibility to inspector error. A more radical approach is to eliminate the inspection function entirely and place quality responsibility in the hands of the people making the product. Although this second approach sounds risky and radical, it usually results in both improved quality and lower cost.

Key Questions

How do we assign inspection points?

How do our returns and warranty claims compare to our final inspection results?

Do we use redundant inspections?

Do any of our receiving inspections duplicate supplier final inspections?

Do our manufacturing people feel responsible for the quality of what they produce?

The Inspection Cost-Reduction Road Map

Inspection adds nothing to the product's value.

Think about that for a moment.

Inspections of incoming material, work in process, and finished products add nothing to the product's value. The only thing inspection does is sort good product from bad and add cost.

This chapter will discuss two approaches for reducing inspection cost:

- The first is to evaluate and improve what the inspectors do and where they do it. This discussion will consider assigning inspection points and modifying inspection methods so that they do a better job of unearthing defects sooner. This will drive the costs of rework, repairs, and scrap lower. We touched on this in an earlier chapter, and we'll explore the topic in more depth here.

- The other approach is far more radical. It eliminates the need for a separate inspection function altogether. This may sound outrageous and risky, but many companies have done this and experienced significant quality improvement.

Improving Inspection

The idea behind inspection is simple. The purpose of an inspection is to sort good product from bad. The inherent inspection assumption is that defects exist. We'll see in the second half of this chapter that this doesn't have to be true, but

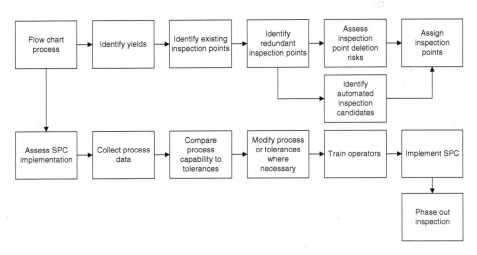

Figure 28.1 Recommended inspection cost-reduction approach.

for now, let's assume that it is. Given that defective products exist, the inspector's mission is to find them so that they can be sorted from the good product.

A question emerges at this point:

Where in the process should the inspections be performed?

The answer to this question is determined by the following factors:

- Inspections should be performed where defects are easy to find and relatively inexpensive to correct. We want to place inspections at these points in the process because if a defect escapes (in other words, if the defect is not detected), it would be expensive and troublesome to correct at some later point. This might be because subsequent process steps make the defect harder to get to, or harder to detect, or both.
- Inspections should be performed where defects are likely to occur. In other words, if the process yield is less than 100 percent (perhaps significantly so), an inspection point will find the defect so that it can be addressed before the product moves further.
- Inspection should be performed prior to the point where defects (if undetected) will have significantly adverse effects. For example, if an expensive warranty repair or an unsafe situation could result from an undetected defect, inspecting the product prior to that point is desirable because it will sort the defective product from the good product.

As a first step in assessing where to place inspection points, it is a good idea to flowchart the process. As explained in an earlier chapter, this is usually a revealing exercise because when flowcharts are prepared, it is often amazing to see just how many inspection points the process contains. When the flowchart shows the process and the inspection points, most people wonder how defects escape at all (given the number of inspections). We'll say more about this in a second.

Once the inspection points are identified, the process yields should be identified at each process step. If the yields at a process step are 100 percent, no inspection is needed at that point unless the consequences of a nonconformance are dire.

In many cases, inspection points can be combined. The concept is that if the product is inspected in two subsequent steps, the first inspection can be eliminated and the product can be inspected only at what was the second inspection point. This is not restricted to two inspection points; in many cases, three or more inspection points can be combined.

In some cases, inspections are redundant. In other words, two inspectors perform the same inspection on the same parts, one after the other. This can result from poor process design by the people assigning the inspection points or by a desire to use redundant inspections to improve quality. Unfortunately, redundant inspections hurt quality. The reason for this is that the inspectors realize their work is being checked by other inspectors, and the motivation to detect defects lessens. Wherever redundant inspections exist, the redundancy should be eliminated.

Another likely spot for redundant inspections is in the receiving inspection area. This is the area in which supplier materials are inspected before being accepted into inventory. There's a good chance the receiving inspections duplicate at least a portion of the supplier's outgoing inspections. A good way to cut costs in this area is to ask for the supplier's final inspection results on a certificate of conformance, and eliminate some or all of the dimensional and other inspections performed in your receiving inspection area. A certificate of conformance is a statement from the supplier that shows the material meets its requirements. Some organizations say they won't rely on a supplier certificate of conformance because they don't trust the supplier's inspection data. If that's the case, our advice is to change suppliers. You can still spot-check the supplier's inspection results, but duplicating an inspection you are already paying for is wasteful.

Another way to cut inspection costs is to modify the sampling approach. In some high-risk areas it might make sense to inspect every part (a 100 percent sampling plan), but in most cases this is overkill. A much smaller sample can be selected and inspected from each production lot. Sample sizes based on lot size and acceptable risk are readily available in quality assurance standards and texts. Using a sampling approach greatly reduces the inspection workload with minimal increased risk of a defect escaping. It's another way to reduce costs.

Automated inspection equipment can increase the accuracy of an inspection and reduce inspection labor content. Approaches for this include dimensional inspection devices that automatically kick out nonconforming parts, scales for weight measurement that do the same, and other similar automated measurement equipment. Coordinate measuring machines can be programmed to perform numerous complex dimensional inspections and print the results. Automated inspection techniques increase inspection accuracy and reduce inspection labor, but they do nothing to reduce the quantity of nonconforming parts. Costs associated with the nonconformances' existence remain.

You might be tempted to delegate the designation of inspection points and sampling plans entirely to the quality assurance or manufacturing engineering department, ignore this as a potential cost-reduction opportunity, and not ask for an analysis of the type described earlier. It's appropriate to trust the judgment of others, but others may not be aware of the potential for cost reductions described here. Inspection costs a lot, so probing the inspection approach makes sense when seeking cost reductions.

A More Radical Approach

The other approach to reducing inspection costs is to eliminate the inspection functions altogether. Many people cringe when they hear this, and the immediate reaction is that product quality will deteriorate. In reality, the opposite occurs. The approach does not eliminate the requirement to make conforming parts and assemblies; it simply shifts the responsibility for assuring the parts are conforming from the inspectors to the people making the product. When this occurs, the inspectors and their costs are no longer necessary.

Inspection is an inefficient process that fails to detect all nonconformances. Numerous studies have shown that under the best of conditions, inspectors will miss 20 percent of the defects. This doesn't mean the product will have a 20 percent defect rate; it just means that whatever the defect rate is, 20 percent of the defects will be missed by the inspectors. This is under ideal conditions, with good lighting and a good work environment. If you doubt this is the case, you can confirm it with a quick experiment. Ask five of your best inspectors to count the number of times the letter "e" appears on this page. It's doubtful you'll get even two answers that are the same.

As mentioned earlier, some organizations attempt to address this by assigning a redundant inspector to reinspect the work of the first inspector. This is expensive and it will have an effect opposite to the one desired. The defect escape rate will actually increase. Both inspectors realize their work is redundant and they become less diligent. The operator (the person making the items being inspected) knows that his or her work is being inspected twice. The operator will also become less diligent, assuming others will catch any nonconformances.

Recognizing this, progressive organizations realize that the answer to improving quality must lie elsewhere. Inspection cannot be counted on for sorting good product from bad. If the objective is to produce products that conform, the organization needs to modify the process such that the yield is 100 percent. If this can be done, the need for inspection and its associated inefficiencies and costs goes away. This may sound like a difficult thing to do, but it is not. Approaches for doing so are described next.

- The organization can (and should) assign teams to find and eliminate the causes of dominant nonconformances, as described in the chapter on scrap and rework reduction.

- Where possible, and as discussed in an earlier chapter, dimensional tolerances can be enlarged consistent with the product's fit and function requirements. If the tolerances can be enlarged so that they are consistent with the process capability, the rejections will drop to zero.

- Assemblies can be designed such that parts are irreversible (d-shaped cut-outs or other indexing features), or they can be designed such that reverse installation has no effect. This will eliminate assembly nonconformances.

- Other design features can be incorporated to prevent improper assembly.

- In areas where it is feasible to do so, the organization can implement statistical process control. Statistical process control is a wonderful approach that analyzes the sources of variability in a production process, minimizes these sources, and then has the *operator* (not a downstream inspector) measure parts as they are produced. These data are graphically posted and monitored in real time, which allows the operator to observe trends as the parts are made. If an adverse trend begins to develop, the operator sees the trend emerging graphically while the parts are still in conformance. The operator can then adjust the process to correct the trend before any defective parts result.

Implementing statistical process control eliminates the need for a separate inspection function (which reduces cost), drives the defect rate to zero (further reducing cost), and places the responsibility for making conforming hardware in the hands of the operator. A detailed description of the statistical process control implementation approach is beyond the scope of this book, but many excellent sources are readily available.[1] The bottom line is that if your organization is not using statistical process control, it should take a hard look at doing so. It is a significant cost-reduction opportunity.

Who should Do This Work

Any initiative about transferring the responsibility for quality from the inspection function to the manufacturing organization has to come from the top. The chief executive has to be behind this or it is not going to happen.

If the chief executive decides that quality responsibility will transfer to the manufacturing group, nearly every group in the company will play a role:

- The manufacturing engineering and engineering groups will have to assure that the drawings' tolerances are consistent with the processes' capabilities.

- The quality assurance group's role will change, and rather than being sorters, they will have to assume different responsibilities. Some will become internal consultants who assist in redesigning processes to eliminate nonconformances. Some will move into manufacturing and make product

---

1. *Quality Management for the Technology Sector*, by this author, is a good source for learning how to implement statistical process control.

instead of inspecting it. Some will become statistical process control experts and assist in statistical process control implementation.

- Purchasing will have to implement certified supplier programs, coordinate obtaining certificates of conformance, and perform other duties to allow elimination of the receiving inspection function.

If the organization opts not to do the aforementioned changes, it should still pursue the cost-reduction opportunities outlined in the first half of this chapter. In that case, the manufacturing engineering and quality assurance groups should take a lead role in redesigning inspection locations and methods.

Risks

There are risks associated with eliminating, combining, or modifying inspections. These risks and associated risk control measures are described next.

- As mentioned earlier, moving from 100 percent inspection to a sampling approach increases the risk that defects will not be detected. This risk can be managed by using sampling approaches defined in quality assurance standards and texts. The quality assurance group will have this information.
- Combining inspections will make the cost of a downstream rejection discovery more expensive, as the product's labor content will be higher. This risk can be minimized by appropriate inspection point selection.
- Eliminating inspection points will present risks similar to those described in the previous bullet. Consideration of the process yield, process improvement, and appropriate inspection point selection will minimize this risk.

There are risks associated with eliminating the inspection function, but surprisingly, the risks are not associated with a reduction in product quality (as might be expected). These are the problems normally encountered, along with suggestions for addressing them.

- Customers may react negatively based on a fear that product quality will decrease. This risk can be managed by product conformance metrics and customer education.
- The operators may view the effort as a way to dump the inspector's duties on them as extra work. Education is the key here.
- Operators may resist the added responsibility for product quality. We're not talking about the aforementioned risk (the resistance to extra work). Here, we are describing a fundamental shift in how responsibilities are allocated. Many people in traditional, inspection-oriented manufacturing environments truly

believe that the operators are responsible for making product, and the inspectors are responsible for product quality. It seems obvious that the responsibility for quality lies in the hands of the people making the product, but a few will still resist. Again, education is the answer to this risk.

- The inspectors will fear losing their jobs. In well-managed companies, the inspectors don't lose their jobs; they are assigned to production roles, thereby increasing the organization's capacity and throughput.

- In union environments, the union will probably resist the approach. The key to managing this risk is to educate the union leadership that the inspectors won't lose their jobs (as described earlier). If the union relationship is managed well, the union can become a strong supporter of this approach, further facilitating its implementation.

References

J. Berk, *Quality Management for the Technology Sector*, Boston, Newnes, 2000.

P.J. Ross, *Taguchi Techniques for Quality Engineering*, New York, McGraw-Hill, 1988.

Part VI

Gaining Disciples and Measuring Progress

We are close to the end of this book. We've reviewed cost-reduction opportunities and approaches in the areas of labor, material, design, process, and overhead. We have two more short chapters and then it's time to get back to work.

The next chapter talks about how to get everyone in the plant thinking about cost reduction through an incentivized cost-reduction program. There's nothing but good news here. The folks in the factory and in the office are smart people with great ideas. Implementing an incentivized suggestion program will focus that energy on finding ways to reduce cost. It will turn everyone into a cost-reduction disciple.

Our final chapter addresses how we measure cost-reduction progress. It offers a simple approach for measuring how well the cost-reduction team and the entire organization are doing.

29

Suggestion Programs

The Bottom Line

Employees in the shop and in the office are closest to the work, and they have great ideas for reducing cost. In many cases, though, employees are hesitant to make improvement suggestions. An incentivized suggestion program will overcome this hesitance, prompt employees to think about cost reduction, and reveal significant opportunities. Successful suggestion programs are objective, responsive, publicized, and financially meaningful.

Key Questions

Do we have a suggestion program, and if so, is it active?

What incentives do our employees have to share their cost-reduction ideas with management?

The Suggestion Program Road Map

An incentivized cost-reduction program is a significant way to get the rest of the organization involved with the cost-reduction program. Many of the best cost-reduction ideas come from the people on the shop floor. They are closest to the work and they often have an intuitive feel for how to do it better. Office workers also have great ideas on how to do their work more efficiently. An incentivized employee suggestion program can tap into this resource.

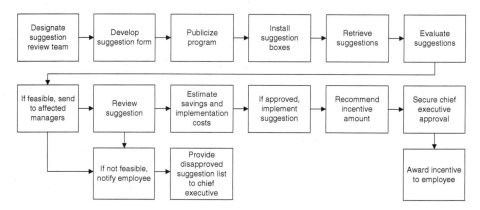

Figure 29.1 Recommended suggestion program approach.

The key to making an employee suggestion program work is to make it simple, make the suggestion evaluations objective, have quick response times, give significant incentives, and publicize the results.

The steps involved in starting and sustaining an employee suggestion program are described next.

- Management should designate an unbiased suggestion review team to take ownership of the program. These programs work best when midlevel managers or supervisors are on the team. People who are on the cost-reduction team are ideal candidates.

- The suggestion review team should develop a simple suggestion form. Simple is the key word here. It's not a good idea to require the people making suggestions to estimate the implementation costs or the savings (that may intimidate some people). All that's needed is the employee's name, the date, and a brief description of the suggestion. Figure 29.2 shows a form that works well.

- The suggestion review team and the organization's management should publicize the suggestion program. Flyers posted near key locations and asking supervisors to explain the program to their employees will help. You don't have to overdo this part or worry about getting it perfect. The best publicity will be when meaningful financial awards go out. We'll get to that shortly.

- The suggestion review team should place suggestion boxes and forms in key locations. If the organization has more than one building, each building should have a suggestion box.

- Someone from the suggestion review team should empty the suggestion boxes on a daily or weekly basis, and provide copies of the suggestions to members of the team.

Suggestion submittal form	
Name (please print):	Date:
Suggestion (describe or draw):	

Figure 29.2 Suggestion form. This simple form can be used for employee suggestions.

- The suggestion review team should meet on a weekly basis to review the suggestions.
- The team should send a note to each person who submitted a suggestion to let that person know the suggestion is in the review cycle, and to thank them for the suggestion.[1]

---

1. Years ago, the U.S. Army at Fort Bliss, Texas, had an incentivized suggestion program similar to the one described here. As soon as the Army received a suggestion and before it was evaluated, tit sent a thank-you note to the soldier who made the suggestion. The note had a new penny glued to it with a caption that said, "A Penny For Your Thoughts." It was a nice touch. The financial awards for approved suggestions were substantially higher.

- If the suggestion review team feels the suggestion has no merit, the team should send a note to the person who made the suggestion to thank them, and to tell them why the suggestion was rejected. The note should be positive, and it should encourage the person who made the suggestion to continue to submit ideas. It's not a bad idea for someone from the team to visit the employee, thank him or her, and explain why the suggestion was rejected.

- If the suggestion review team feels the suggestion has merit, they should send it to the department managers who have expertise in the areas addressed by the suggestion. This will be a request to the department managers to review and approve or disapprove the suggestion. Figure 29.3 shows a recommended form.

- The department managers should quantify the annualized savings, the implementation cost, and any associated implementation risks. If the managers agree with the suggestion, they should implement it.

- If the managers feel the suggestion is not workable, they should explain why. Two weeks should be enough time to respond to each suggestion, although the suggestion review team can modify the allowed response time where it makes sense to do so. The point is that there should be a required response due within a reasonable period. The suggestions can't be allowed to fall into a black hole.

- On approved suggestions, the suggestion review team should recommend an incentive amount based on the annualized savings. Although the amount is discretionary, one percent of the total annualized savings is usually enough to make the incentives meaningful.

- The suggestion review team should meet with the organization's senior management to negotiate the incentive amount. It won't be a tough negotiation. Chief executives love doing this.

- On a monthly basis (or more often if desired), the organization's chief executive should visit employees whose suggestions have been approved to present the incentive checks. Some organizations prefer to do this at a meeting with all of the company's employees. Others prefer to do it in small meetings in the work center. Either approach works well. Word of the incentives will travel quickly.

- On a monthly basis, the suggestion review team should send a summary of all rejected suggestions to the chief executive, along with the reasons for the rejection.

Who Should Do This Work

As indicated earlier, suggestion review team participation should be an additional duty for the existing staff. It won't take much time, and the people who are on these committees usually enjoy their participation.

Suggestion evaluation form

Suggestion evaluation form (please see attached suggestion)	
To:	From:
Date:	Response due date:
Approval decision Approved ☐ Disapproved ☐	
Rationale:	
Predicted implementation cost and basis:	
Predicted annual cost savings and basis:	
Reviewer name:	Date:

Figure 29.3 Suggestion review form. The suggestion review team can transmit cost-reduction suggestions (using this form) to the department manager(s).

The suggestion review team will not be able to direct implementation of good suggestions. That will have to come from the department managers and in some cases, if the implementation costs are significant, from the chief executive.

Risks

In some cases the "not invented here" syndrome may surface, and suggestions that the review team feels are acceptable may be rejected by the managers who will have to implement them. Sometimes the reasons for rejection are legitimate, and sometimes they are not. Managers may be embarrassed because they didn't have the idea first. Managers may reject the idea because they feel it infringes on their authority. Sometimes it might be personal, and the manager just doesn't like the person who made the suggestion. All of these risks can be mitigated by remaining objective and attempting to understand the reasons for the rejection.. If the idea is a good one and the affected managers cannot be persuaded, the chief executive's monthly rejected suggestion summary may result in reconsideration. The good news is that most managers are extremely receptive to ideas that reduce cost.

On rare occasions, someone other than the person who submitted the suggestion may claim that a submitted idea was his or hers. This risk can be mitigated by requiring a date on the suggestion form, and by explaining to everyone that in order to be considered for an incentive award, an idea must be submitted through the suggestion program.

On some occasions, concerns may arise about suggestions coming from people paid to take cost out of products and processes. For example, manufacturing engineers are paid to design efficient production processes. If a manufacturing engineer submits a process improvement idea, should they receive additional money for it? There is no right or wrong answer to this question. Some companies accept the suggestions and pay incentives for the ones they approve, and some feel that suggestions coming from people paid to improve efficiency (such as manufacturing engineers) are not eligible for the incentive. In our experience, these programs seem to work best if everyone is eligible.

References

P. Townsend, and J. Gebhardt, *The Executive Guide to Understanding and Implementing Employee Engagement Programs: Expand Production Capacity, Increase Revenue, and Save Jobs*, Milwaukee, American Society for Quality, Quality Press, 2007.

B. Nelson, *1001 Ways to Energize Employees*, Markham, Canada, Thomas Allen and Son, Ltd., 1997

30

Measuring Progress

Table 30.1 summarizes the metrics described throughout this book and suggests how they can be used to measure cost-reduction progress. Many of the techniques summarized here are included in the free Excel template described throughout this book (please visit www.ManufacturingTraining.com to download it).

Table 30.1 Tools for Measuring Cost Reduction

Tool	Description	Suggested Review Frequency
Overall cost-reduction plan	Action table defining actions, assignees, required completion dates, estimated annual savings	Weekly
Headcount	X-Y plot showing authorized direct labor headcount versus actual headcount, and authorized overhead headcount versus actual headcount	Weekly for direct labor work center supervisors; Monthly for middle and senior management
Efficiency	X-Y plot showing individual, work center, and plant efficiencies and efficiency trends	Daily and weekly for work center supervisors; Monthly for middle and senior management
Machine utilization	X-Y plot showing machine, work center, and plant utilization and utilization trends	Weekly for work center supervisors; Monthly for middle and senior management

Table 30.1 (cont.) Tools for Measuring Cost Reduction

Tool	Description	Suggested Review Frequency
Personnel utilization	Percent of time charged to overhead versus direct labor at the individual, work center, and plant levels	Weekly for work center supervisors; Monthly for middle and senior management
Lost time	Delay ratio analysis by work center and plant	Monthly
Overtime	X-Y plot showing overtime budget versus actual overtime	Daily for work center supervisors; Weekly for middle management; Monthly for senior management
Learning curve	X-Y plot showing predicted learning curve for each product line and the organization's compliance to it	Quarterly
Material utilization	Paint consumption and thickness, materials consumption normalized to production rates, scrap sales	Daily for work center supervisors; Weekly for middle management; Monthly for senior management
Inventory turns	Inventory turns calculation	Monthly
Inventory value	Total inventory value calculation	Monthly
Scrap, rework, and repair cost	Pareto charts, and trend charts showing costs of scrap, rework, and repair	Monthly
Plant cleanliness	Weekly plant cleanliness tracking board review, plant tours	Weekly
Overhead	X-Y plot showing plan versus actual overhead costs, with three-month look-ahead by each manager	Monthly
Approved suggestions	Running tally of cumulative cost savings and cumulative quantity approved	Monthly
Suggestion program rejected suggestions	List summarizing rejected suggestions	Monthly for middle and senior management
Financial reviews	Normal organization financial reviews, showing sales, profit, cost of goods sold, etc.	Monthly

In the final analysis, the best measure of an organization's overall success in controlling cost is its income statement. The intent of this book (and the technologies it contains) is to outline the actions required to help companies maximize their profits. We wish you every success in doing so.

Index